PROGRESS IN MATERIALS SCIENCE, VOLUME 11

(*INCORPORATING PROGRESS IN METAL PHYSICS*)

EDITOR: BRUCE CHALMERS

CONDENSATION AND EVAPORATION

Nucleation and Growth Kinetics

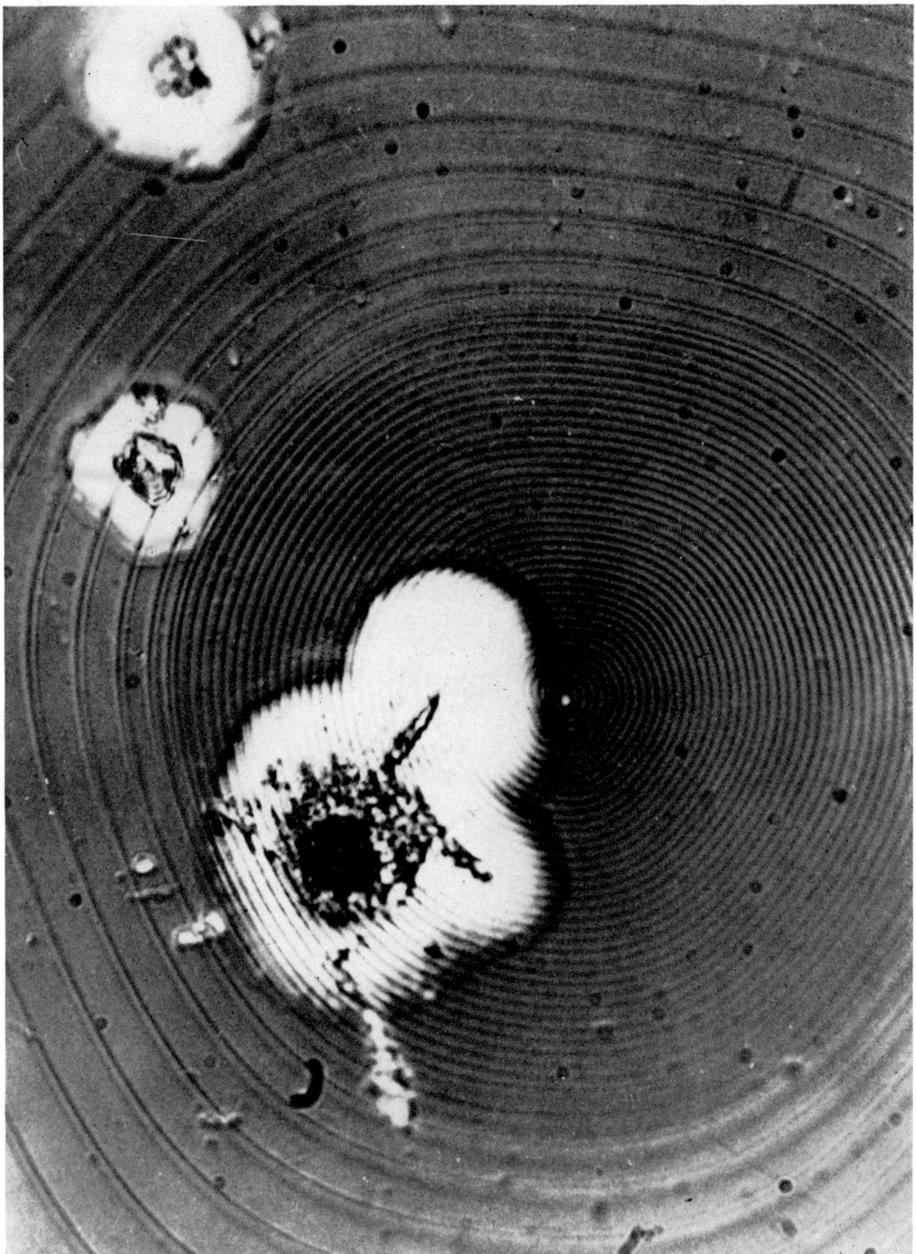

Growth spiral at point of emergence of spiral dislocation on surface of silicon carbide.[525]
1500× (reduced to 7/10ths).

CONDENSATION AND EVAPORATION

Nucleation and Growth Kinetics

BY

J. P. HIRTH

Associate Professor of Metallurgical Engineering
The Ohio State University

and

G. M. POUND

Alcoa Professor of Light Metals
Carnegie Institute of Technology

PERGAMON PRESS

OXFORD · LONDON · PARIS · FRANKFURT

1963

PERGAMON PRESS LTD.
Headington Hill Hall, Oxford
4 & 5 *Fitzroy Square, London, W.*1

GAUTHIER-VILLARS ED.
55 *Quai des Grands-Augustins, Paris* 6e

PERGAMON PRESS G.m.b.H.
Kaiserstrasse 75, *Frankfurt am Main*

Distributed in the Western Hemisphere by
THE MACMILLAN COMPANY · NEW YORK
pursuant to a special arrangement with
Pergamon Press Limited

Library of Congress Catalog Card Number 49–50107

PRINTED IN GREAT BRITAIN BY THE PITMAN PRESS, BATH

CONTENTS

FOREWORD

THIS description of phase transformations involving the vapor phase, including the kinetics of nucleation and growth in such transformations, is written with the purpose of providing a rather complete review of theory and experiment in the field. An attempt is made to point out areas where theory and experiment are not in agreement and where further research should be fruitful. Also, this book is intended to be a reference textbook for a graduate or advanced undergraduate course on the kinetics of phase transformations. In fact, we have used much of the following material as a basis for such courses at our respective schools.

We expect that eventually the theory of these transformations will be couched in quantum mechanics, involving considerations of surface states, interatomic forces, quantum mechanics of irreversible processes, etc. At present these problems still await solution and in this work we pursue an essentially classical description of the processes of interest here, although in some cases we discuss the limitations of such a description.

The principal background concepts that we use are: the kinetic theory of gases (e.g. reference 1); the description of a surface as being made up of atomic or molecular building blocks;[2,3,4] the principle of microscopic reversibility;[5] statistical thermodynamics; and the precepts of absolute rate theory.[6]

In rationalizing the results of experiments, the data in many cases are not sufficiently complete to yield an unequivocal interpretation. In such cases we have occasionally indulged in speculation, but we have tried to label these speculations clearly.

In the following, then, we consider in succession: thermal accommodation and the relation of thermal accommodation to the probability that a molecule incident onto a surface will be adsorbed; homogeneous and heterogeneous nucleation of condensed phases from the vapor; heterogeneous nucleation upon substrates; growth and evaporation of liquids and perfect crystals; growth and evaporation of imperfect crystals; nucleation in ebullition and cavitation of liquids; and finally, void formation in solids.

ACKNOWLEDGEMENTS

ONE of the authors (J.P.H.) gratefully acknowledges the support of the U.S. Office of Naval Research, the other (G.M.P.) that of the U.S. Air Force Office of Scientific Research and the U.S. Air Force Geophysical Research Laboratory. Indeed, these agencies sponsored an appreciable fraction of the work described herein.

The authors thank their colleagues in the Metals Research Laboratory and the Department of Metallurgical Engineering at the Carnegie Institute of Technology, in particular Professors W. R. Bitler and J. Lothe, for illuminating discussions. Also, thanks are due to Dr. J. H. McFee of the Bell Telephone Research Laboratories and Dr. R. A. Rapp of the Wright Air Development Center for helpful criticisms of the manuscript. The authors are especially pleased to acknowledge the large contributions of the doctoral researches of Drs. J. Chirigos, K. L. Moazed, R. A. Rapp and J. H. McFee, all recently of the Carnegie Institute of Technology, and Messrs. B. K. Chakraverty and S. J. Hruska. The authors are indebted to Drs. W. A. Chupka of the Argonne National Laboratory, E. von Goeler and E. Luscher of the University of Illinois, U. Merten of the General Atomics Research Laboratory, G. W. Sears of the Stromberg-Carlson Division of the General Dynamics Corporation, R. L. Parker of the National Bureau of Standards, B. M. Siegel of Cornell University, A. J. Forty of Bristol University, P. B. Price of the General Electric Research Laboratory, and Professor J. L. Margrave of the University of Wisconsin for results of unpublished research. They thank Dr. W. C. Dash of the General Electric Research Laboratory for contributing Fig. 45, Dr. I. M. Dawson of the University of Glasgow for Figs. 41 and 42, and Dr. F. H. Horn of the General Electric Research Laboratory for the Frontispiece. Finally, appreciation is expressed to Professor Dr. Robert F. Mehl, formerly Dean of Graduate Studies, Head of the Department of Metallurgical Engineering, and Director of the Metals Research Laboratory at the Carnegie Institute of Technology and currently Consultant to the U.S. Steel Corporation in Zürich, for inspiring the writers to undertake this work.

LIST OF NOMENCLATURE

a	= jump distance on surface, lattice constant.
A	= area.
A_1	= fraction of the evaporating area near ledge sources.
A_2	= fraction of the evaporating area where $\lambda \simeq \lambda_0$ and nearly asymptotic ledge motion obtains.
A	= quantity related to relaxation time for thermal equilibration.
α_c	= general condensation coefficient.
α_{c_i}	= specific condensation coefficient for specific constraint.
α_λ	= evaporation coefficient for surface-diffusion constraint.
α_Υ	= evaporation coefficient for surface-diffusion constraint.
α_T	= thermal accommodation coefficient.
α_v	= evaporation coefficient, general.
α_{v_i}	= specific evaporation coefficient for specific constraint.
b	= magnitude of the Burgers vector, platelet thickness.
B	= relatively temperature-insensitive pre-exponential of nucleation rate expression.
B_i	= cluster containing i particles.
β	= sticking coefficient.
c	= mole fraction of first component in the critical nucleus.
C	= whisker circumference.
C_p	= molar heat capacity at constant pressure.
C_v	= molar heat capacity at constant volume.
C_1'	= relatively temperature-insensitive factor in pre-exponential of nucleation rate expression.
D	= general diffusion coefficient.
D_s	= surface diffusion coefficient of adsorbed molecules on substrate.
D_L	= diffusion coefficient for ledge.
D_v	= diffusion coefficient in vapor phase.
$\bar{\delta}$	= "width" of the top of the activation barrier in the direction of decomposition.
δ	= "free angle" ratio.
$\delta_{\mathrm{k}-1}$	= "free angle" ratio for kink-ledge kinetics.
$\delta_{1-\mathrm{ad}}$	= "free angle" ratio for ledge-adsorbed kinetics.
δ_{ad}	= "free angle" ratio for adsorption from the vapor.
δ'	= displacement of the surface of tension beneath the surface of discontinuity.
Δ	= distance at which steady-state concentration is maintained.
e	= electronic charge.
E_{adion}	= potential energy of desorption for ions.

E_{adatom} = potential energy of desorption for atoms.
E_{ion} = ionization potential of condensate atoms.
E_I = kinetic energy of incident beam.
E_R = kinetic energy of rebounding molecules.
E_s = kinetic energy of emergent molecules which have equilibrated with substrate.
E_W = work function of substrate.
$E_{0_{\text{des}}}$ = potential energy of desorption.
$\Delta E_{0_{\text{vap}}}$ = potential energy difference between reactant and activated states in vaporization.
ΔE = energy of evaporation.
E_{el} = elastic strain energy which was in liquid.
ε = edge energy of disc.
ε' = bond energy.
ε_{c-v} = crystal-vapor edge energy.
$\mathscr{E}$ = dielectric constant.
$\Delta f'$ = free energy of material of the local density or composition in the absence of a gradient.
$f_R^\star$ = rotational partition function in the activated liquid state.
f_{R_L} = rotational partition function in the restricted liquid state.
f_{R_v} = rotational partition function in the vapor phase.
$f_{\text{tr}}^\star$ = translational partition function for the direction in which evaporation is occurring.
f_V = vibrational partition function.
F = partition function for the reactant state.
$F^\star$ = partition function for the activated state.
F_{vap} = partition function in the vapor.
g = number of independent jump directions on the surface, gradient of density or composition in the X direction normal to the interface.
ΔG^0 = Gibbs' free energy of formation.
$\Delta G_{\text{des}}^\star$ = Gibbs' free energy of activation for desorption.
$\Delta G^\star$ = free energy of formation of a critical nucleus at rest.
ΔG_s = free energy change accompanying separation of a group of molecules from a larger ensemble.
ΔG_r = free energy change accompanying activation of rotational degrees of freedom.
ΔG_t = free energy change accompanying activation of translational degrees of freedom.
G_A = free energy of crystallite due to presence of a surface.
ΔG_q = free energy change accompanying distribution of nuclei on a substrate having q sites per unit area.
$\Delta G_{\text{vd}}^\star$ = activational free energy for vacancy diffusion.
$\Delta G_\perp^\star$ = free energy of formation if a critical nucleus forms an incoherent interface with substrate.

$\Delta G_L^\star$ = free energy of formation of a critical nucleus at a macroscopic ledge.

$\Delta G_{\text{vap}}^\star$ = Gibbs free energy of activation for evaporation.

$\Delta G_{\text{hd}}^\star$ = activational free energy for the jump of a vacancy near the surface.

$\Delta G_{\text{d}}^\star$ = free energy of activation for diffusion in the liquid.

$\Delta G_{\text{sd}}^\star$ = free energy of activation for a surface-diffusion jump.

ΔG_v = Gibbs free energy change per unit volume of product material.

h_I = step height of asperities.

ΔH_{vap} = enthalpy of vaporization.

ΔH_{hole} = quantity reflecting the probability that a "hole" exists in the liquid surface.

$\Delta H_{L_1}^\star$ = enthalpy of activation for the step liquid → adsorbed.

$\Delta H_{L_2}^\star$ = enthalpy of activation for the step adsorbed → liquid.

$\Delta H_{L_3}^\star$ = enthalpy of activation for the step adsorbed → vapor.

$\Delta H_{\text{k}-1}^\star$ = enthalpy of activation for an atom to move from a kink to a position at a ledge.

$\Delta H_{\text{ld}}^\star$ = enthalpy of activation for diffusion at a ledge.

$\Delta H_{\text{l}-\text{ad}}^\star$ = enthalpy of activation for an atom to dissociate from a ledge to an adsorbed position.

$\Delta H_{\text{sd}}^\star$ = enthalpy of activation for surface diffusion.

ΔH_0° = standard enthalpy of formation at absolute zero.

i = number of molecules.

$i^\star$ = number of molecules in critical nucleus.

$i_{\text{vac}}^\star$ = number of vacancies (or primitive units of volume increase) in the critical nucleus.

I = moment of inertia.

$\jmath$ = integral number of molecular ledge units in a macroscopic step.

J_c = net condensation flux.

J_{c_g} = gross condensation flux.

J_v = net vaporization flux.

J_{v_g} = gross evaporation flux.

J = nucleation rate.

J_L = nucleation rate at macroscopic ledge.

J_{L_1} = flux from liquid → adsorbed states.

J_{L_2} = flux from adsorbed → liquid states.

J_{L_3} = flux from adsorbed → vapor states.

J_{L_4} = flux from vapor → adsorbed states.

$J_\perp$ = rate of formation of nuclei which form incoherent interfaces with substrate, nucleation rate at dislocations.

J_{imp} = rate of impurity adsorption.

J_{tip} = net impingement flux at whisker tip.

k = Boltzmann's constant.

K = constant determined principally by the intermolecular forces in the particular system.

K_1 = constant characteristic for nucleation by ions.
K_2 = constant characteristic for nucleation in thermal spikes.
l = whisker length.
$\dot{l}$ = whisker growth rate.
L = approximately constant quantity.
λ_I = de Broglie wave length of the incident molecular beam.
λ = interledge spacing.
Γ = Lothe–Pound correction.
m = molecular mass.
μ_c = chemical potential of molecules in the nucleus.
μ_v = chemical potential of molecules in the vapor.
μ = shear modulus.
n_i = concentration of clusters containing i molecules.
n_1 = concentration of monomer.
n_c = concentration of ions.
$\bar{n}_{c_0}$ = steady-state concentration of ions in air in the absence of a field.
$\dot{n}_c$ = rate of production of ions.
$\bar{n}_c$ = steady-state concentration of ions in a field.
n_m = maximum concentration of droplets.
n_0 = concentration of monomer in the liquid.
n_{s_e} = adatom population in equilibrium with the vapor of bulk condensate.
n_s = adatom population.
n_s' = integral condensate in a given time, population in equilibrium with a monatomic ledge.
$n^\star$ = concentration of atoms in the activated state.
n_L = concentration of atoms in the reactant state, or in a liquid surface.
$n_{s_{\text{ion}}}$ = concentration of adions.
n_v = concentration of the evaporating substance in the vapor phase.
n_{h_e} = equilibrium concentration of surface vacancies.
n_h = concentration of surface vacancies.
n_{K} = number of kinks per unit length.
n_{L} = concentration of molecules at a ledge.
n_{L}' = concentration of molecules at a ledge and immediately adjacent to a kink.
n_{vac} = actual concentration of vacancies.
n_{vac_e} = equilibrium concentration of vacancies.
N = Avogadro's number; concentration of droplets.
η = angle between planes forming a macroscopic step.
ω = frequency with which a single atom joins a critical nucleus to promote it to a stable growing nucleus.
$\bar{\omega}$ = frequency factor or transmission frequency for the activated species.
Ω = molecular volume.
p = vapor pressure.

p_e = equilibrium vapor pressure.
p_{inert} = pressure of inert gas.
$p^\star$ = partial pressure of vapor in the critical bubble.
P = hydrostatic stress imposed on the liquid.
P' = minimum pressure for bubble formation as calculated from the van der Waals equation.
ϕ_1 and ϕ_2 = appropriate geometric functions of θ.
Ψ = spacing between kinks.
$\Psi(t)$ = factor accounting for induction period.
r = embryo radius or radius of curvature of the spherical bubble.
$r^\star$ = radius of critical nucleus or bubble.
r_0 = limiting hole radius for strain energy effect.
R = gas constant; radius.
ξ = a factor equal to unity for a screw dislocation, one minus Poisson's ratio for an edge dislocation.
ρ_{vap} = density of vapor.
ρ_c = critical radius of curvature.
s = molecular entropy of liquid.
S = supersaturation.
σ = specific interfacial free energy or surface tension.
σ_0 = surface tension at zero curvature.
σ_{x-v} = substrate-vapor specific interfacial free energy.
σ_{c-x} = condensate-substrate specific interfacial free energy.
σ_{c-v} = condensate-vapor specific interfacial free energy.
$\sigma_{\perp}$ = interfacial free energy of incoherent interface.
σ'' = free energy of interface between nucleus and substrate with entrapped adsorbate layer.
t = time.
T = absolute temperature.
T_I = r.m.s. temperature of the incident molecules.
T_R = r.m.s. temperature of the reflected (or re-evaporated) molecules.
T_s = substrate temperature.
T_v = r.m.s. temperature of the evaporating vapor.
T_C = critical temperature of the liquid.
T_Q = temperature of quench.
τ = reciprocal of the velocity $\dot{x}$ of the charged clusters in the applied field.
τ_s = mean residence time of an adatom.
τ_r = relaxation time for thermal equilibration.
τ_k = induction period to establish steady-state.
τ_ν = mean period for atomic jump processes on the surface.
θ = equlibrium contact angle.
θ_I = angle of incidence.
v = r.m.s. velocity or macroscopic step velocity.
v_x = velocity of kinks.

v_ν = velocity of ledges.
V = volume.
$V(r)$ = interaction potential.
ν = vibrational frequency.
ν_0 = Debye frequency.
w = width of the interface.
W = number of complexions; work.
$\dot{x}$ = mean velocity of the activated species in the direction of decomposition.
X = normal distance from a substrate.
$\bar{X}$ = r.m.s. diffusion distance.
$X_\perp$ = fraction of surface sites intersected by dislocations.
y = distance from the platelet center to the midpoint of a growing edge.
$\bar{Y}$ = mean free path of diffusion on a ledge.
z_l = molecular coordination number in the interior of the liquid.
z_s = molecular coordination number in a plane liquid–vapor interface.
Z = Zeldovich non-equilibrium factor.

A. COEFFICIENTS OF CONDENSATION, EVAPORATION AND THERMAL ACCOMMODATION

IN growth of condensed phases from the vapor a prime consideration is the interaction of the impinging molecule with the surface of the condensed phase. The general condensation coefficient α_c is the probability that such an impinging molecule will be adsorbed, thermally equilibrated and incorporated into* the surface in question. Similarly, a closely related quantity, the evaporation coefficient α_v, describes the vaporization of condensed phases. Historically, these factors have been connected to each other and to the thermal accommodation coefficient and the extent of compliance with the cosine law for diffuse reflection in an inconsistent manner. It is the purpose of this Section to make clear the true relationships between these quantities and in particular to show the extent to which data on thermal accommodation and angles of ejection of molecules from surfaces may be used to predict coefficients of condensation and evaporation.

1. CONDENSATION AND EVAPORATION

From the kinetic theory of gases[1] the flux of molecules of a given species striking a surface which is equilibrated with the vapor phase is

$$J_{c_g} = p/(2\pi mkT)^{\frac{1}{2}} \tag{A-1}$$

where p is the vapor pressure of the substance in question, m is the molecular mass, k is Boltzmann's constant and T is the absolute temperature. At equilibrium the net flux is zero and thus the gross evaporation flux J_{v_g} must equal the gross condensation flux J_{c_g}. If there is no coupling between the processes of condensation and evaporation J_{v_g} will equal the gross vaporization flux from the surface in equilibrium with the vapor at surface temperature T and this flux will be independent of the gross condensation flux. Hence in such a situation one obtains by difference the net condensation flux, i.e. the ideal growth equation of Hertz[7] and Knudsen[8]

$$J_c = (p - p_e)/(2\pi mkT)^{\frac{1}{2}} \tag{A-2}$$

where p_e is the equilibrium vapor pressure at surface temperature T and p is the *equivalent vapor pressure* at surface temperature T that would give the gross condensation flux as determined by the experimental conditions at

* At least temporarily.

hand. Thus if the impinging vapor is at a temperature $T' > T$, p as defined in this manner is a fraction $(T/T')^{\frac{1}{2}}$ of the true pressure of the impinging vapor. In other words it is convenient here to define the pressure in terms of a flux of molecules rather than a rate of change of momentum.

Deviations from the ideal equation (A-2) are taken into account by the introduction of a coefficient which has been variously called the condensation coefficient, the vaporization (or evaporation) coefficient and the accommodation coefficient. As will be shown in the following, the coefficient multiplying equation (A-2) depends on whether net vaporization or condensation is occurring. Furthermore the term accommodation coefficient usually refers to the thermal accommodation coefficient, which does not in general bear any direct relation to equation (A-2). For cases in which $p < p_e$ and net vaporization is occurring, we define the general, i.e. complete, evaporation coefficient α_v by the expression for the net vaporization flux

$$J_v = \alpha_v(p_e - p)/(2\pi mkT)^{\frac{1}{2}}. \tag{A-3}$$

Similarly the general condensation coefficient for cases in which $p > p_e$ is given by

$$J_c = \alpha_c(p - p_e)/(2\pi mkT)^{\frac{1}{2}}. \tag{A-4}$$

Furthermore, in many cases more than one constraint* may contribute to a value of α_c or α_v less than unity. Therefore for a specific constraint we define a specific coefficient, for example α_{c_1}, α_{c_2}, etc. Thus the complete condensation (or evaporation) coefficient may be a multiple of several specific coefficients. One notes that for some specific constraints (but not for all) α_{c_i} may equal α_{v_i}. The actual values of α_{c_1} and α_{v_1} are, in our terminology, related to the energetics of interaction of the molecules with the surface and the probability of elastic reflection, and this subject will be discussed in the following. The quantity $(1 - \alpha_{c_1})$ is the fraction of molecules striking the surface which returns to the vapor on elastic rebound from collision. It is evident that $\alpha_{c_1} = \alpha_{v_1}$; if at equilibrium the fraction $(1 - \alpha_{c_1})$ is elastically reflected, α_{c_1} is equal to the fraction of the gross vaporization flux due to actual evaporation and this is the fraction α_{v_1} of the gross equilibrium vaporization flux that would be realized upon evaporation into a vacuum.† Review articles which relate to the problem of molecule-surface interactions have been written by Massey and Burhop,(11) Devienne,(12) Schafer,(13) Herzfeld(14) and Ehrlich.(15)

2. Thermal Accommodation

(a) *Definition of Thermal Accommodation*

The term thermal accommodation was introduced by Knudsen(9) to describe the degree of thermal equilibration with the substrate surface of

* By constraints we mean natural impediments to the process in question.

† Neglecting other constraints.

molecules incident onto and subsequently ejected from it. If T_I and T_R are the r.m.s. temperatures (equal to $mv^2/3k$ where m is molecular mass and v is r.m.s. velocity) of the incident and reflected (or re-evaporated) molecules, respectively, and T_s is the substrate temperature, thermal accommodation is described by the thermal accommodation coefficient*

$$\alpha_T = (T_I - T_R)/(T_I - T_s). \tag{A-5}$$

An equivalent definition is the statement of accommodation coefficient in terms of energies

$$\alpha_T = (E_I - E_R)/(E_I - E_s). \tag{A-6}$$

The problem of thermal accommodation is only indirectly related to deposition processes. However, consideration of thermal accommodation is of interest in that, hypothetically, incomplete thermal accommodation of a molecule impinging on a substrate can be associated with reflection of the molecule and hence with a condensation coefficient less than unity. Two types of vapor–substrate systems are pertinent to thermal accommodation: one where a substrate is exposed to an equilibrated vapor phase and another in which a directed beam of molecules is projected onto a substrate. To illustrate the latter system, which is more important because most experiments on thermal accommodation are performed with directed molecular beams, Fig. 1 shows the interaction of a molecule with a substrate. A normally incident molecule with initial kinetic energy E_I approaches the substrate, is accelerated, suffers collision with the surface and rebounds. Two possibilities for adsorption are: first, if an amount of energy greater than E_I is lost to the surface on collision, that is if the molecule suffers transition into a bound state as shown in Fig. 1, the molecule will not retain sufficient energy to escape on rebound and will oscillate in the potential well; second, even without energy loss, momentum transfer to a velocity component parallel to the surface can reduce the normal component of E by more than E_I, also leading to adsorption of the molecule. If the molecule escapes back to the vapor phase on rebound it is considered to be reflected.

(*b*) *Theory*

Energy exchange and transition probabilities in the case of an incident molecule which interacts with the surface only through Van der Waals forces have been treated in detail by Lennard-Jones and Devonshire,[35] Zener[36] and others, as reviewed recently by Massey and Burhop.[11] Of more interest in the present application to crystal nucleation and growth is the problem of energy exchange and transition probability in the presence of strong

* In Section B on Nucleation of Condensed Phases from the Vapor another kind of thermal accommodation, which has to do with dissipation of the heat of condensation in small clusters of molecules, will be briefly considered.

attractive forces at the surface. The case generally treated is the simple type of interaction illustrated by case (1) in Fig. 1. Langmuir[37] presented qualitative arguments to show that the time of collision of an incident molecule was sufficient for a large fraction of its energy to be dissipated thermally. Hence he concluded that all incident molecules equilibrate with the surface when strong attractive forces are present. Cabrera[38] and Zwanzig[39] have recently considered the problem. They pointed out the inadequacy of the

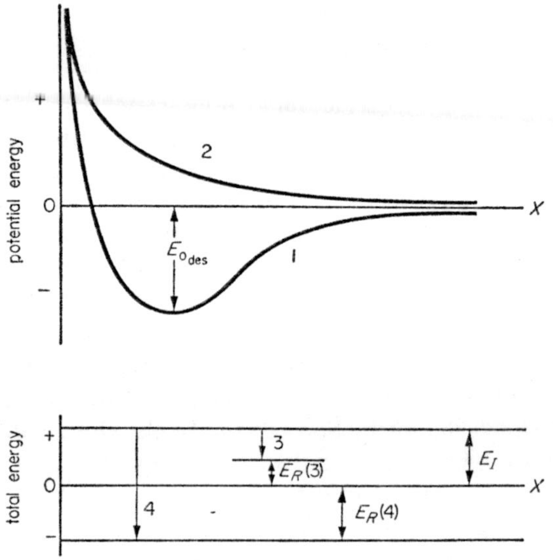

FIG. 1. Rest potential energy of interaction as a function of normal distance X from substrate of a molecule in the case of (1) long-range attraction and short-range repulsion and (2) long and short-range repulsion, and total energy as a function of X showing (3) transfer into a lower energy level which nevertheless permits escape of the molecule on rebound and (4) transfer into a bound state. E_I is the incident kinetic energy of the molecule, E_R the energy retained on rebound. Transfer into a bound state is possible in case (1), not in case (2).

Lennard-Jones one-phonon activation model[35] to explain interaction in the presence of strong attractive forces. Further, they noted the difficulty in treating a multiple-phonon process and instead used a classical approximation. Their model is a one-dimensional semi-infinite array of spring-connected masses impinged on by a gaseous atom with interaction potential $V(r)$. Using Schrödinger's solution[40] for motion of such a chain and supposing that, as a function of displacement from the chain end, the gaseous atom moves in a parabolic potential well that is "cut-off" at the limit of the range of surface forces, Cabrera[38] showed that the atom will lose sufficient energy to the lattice to be captured if the incident energy E_I is equal or less than 25 times the potential energy of desorption $E_{0_{des}}$. Zwanzig[39] verified Cabrera's result and treated several other possible interaction potentials with similar results. This classical model may not be appropriate for incident particles of very low energy which would interact only with phonons of long wave length. The density of these is quite small and one might expect reflection due to a low probability of phonon excitation.

For metals with a monatomic vapor phase, $E_{0_{des}}$ is of the order of 20 to 80 kcal, i.e. the energies of desorption are high. Thus by the above model adsorption will occur unless the beam temperature is of the order of a million degrees, a condition unlikely to be attained in experiment. Only in the presence of a strong impurity adsorbate which could markedly lower $E_{0_{des}}$ is reflection of an incident atom likely.

For non-metals or metals with polymeric vapor species, activational enthalpies or entropies for condensation may be appreciable so that the interaction shown in case (1) of Fig. 1 may no longer occur. However, again under the condition of equilibration of substrate and vapor, an extension[41] of an argument of Millikan[19] which will be stated in the following, shows that α_T approaches unity as equilibrium is approached. This is a consequence of the second law of thermodynamics. Under non-equilibrium conditions an activation energy for adsorption, as postulated by Miyamoto,[42] could lead to reflection of the incident atoms of lowest energy and a thermal accommodation coefficient different from unity.* For example, consider Fig. 2 in

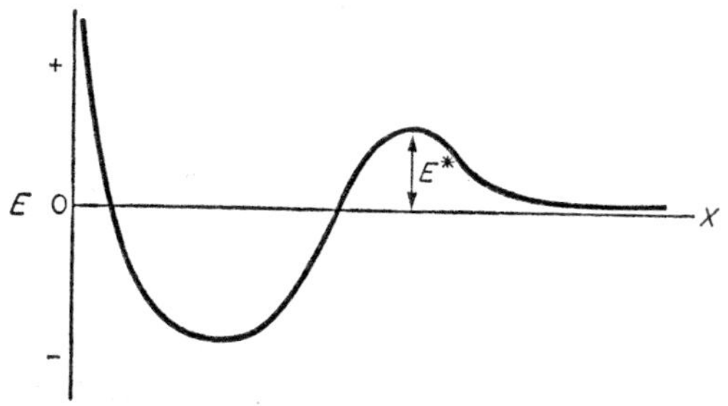

FIG. 2. Rest potential energy of interaction of a molecule as function of X, the normal distance from a substrate.

which incident molecules with energy less than $E^\star$ will be reflected back to the vapor with energy essentially equal to their initial kinetic energy. The portion of the incident atoms that interact with the surface and subsequently re-evaporate will have kinetic energy $E^\star$ in excess of that expected for a vapor in equilibrium with the substrate. Hence for free evaporation in the latter case, the evaporating molecules will have an apparent "temperature" greater

* Fuchs[43] rejects this possibility on the grounds that energies of evaporated atoms generally correspond to surface temperatures. Limited experimental evidence, however, certainly does not seem sufficient to rule out the possibility of an activation energy for condensation.

than the substrate temperature. In fact we may define the equivalent thermal accommodation coefficient for free evaporation

$$\alpha'_T = T_v/T_S \tag{A-7}$$

where T_v is the r.m.s. temperature of the evaporating vapor. This expression is of interest in this discussion because the value of α'_T is related to α_T and to the shape of the potential curve in Fig. 1. That is, a value of $\alpha'_T = 1$ implies that an activation barrier for evaporation is absent and hence that α_T cannot differ from unity because of reflection of impinging molecules that cannot surmount an activational enthalpy barrier for condensation. However, entropy constraints can still lead to $\alpha_T < 1$ even though $\alpha'_T = 1$. An activational entropy for condensation is usually associated with a marked dependence of the interaction potential upon the orientation of the incident molecule so that if improperly oriented the incident molecule may experience only repulsive forces and reflect with $\alpha_T < 1$. For example in Fig. 1, suppose that half the incident molecules were oriented to sense potential (1) while half were oriented to sense potential (2). In this case the latter half would reflect and α_T would be less than unity.*

Finally there exists the possibility that an adsorbed molecule may re-evaporate before equilibrating thermally with the surface, a case that we call quasi-adsorption. After thermal equilibration the residence time will be†

$$\tau_s = (h/kT)[1 - \exp(-h\nu/kT)]^{-1} \exp(\Delta G^\star_{\text{des}}/kT), \tag{A-8}$$

where ν is the surface vibrational frequency. However, before thermal equilibration the frequency of escape will exceed the reciprocal of equation (A-8) by a factor $\exp(A/tkT)$ where A is related to the relaxation time for thermal equilibration τ_r. Hence a large τ_s is not sufficient to predict a thermal accommodation coefficient of unity, but a large τ_s together with a τ_r less than or equal to the period for a few oscillations of a surface atom is sufficient. Now for the case of strong attractive forces Lennard-Jones[44] (using Schrödinger's[40] infinite-chain model) and McFee[45] (using Cabrera's[38] force model of a semi-infinite chain) both estimate $\tau_r \leq 2/\nu_0$ in which ν_0 is the Debye frequency. Thus the condition for thermal equilibration is fulfilled. Lennard-Jones and Strachan[46] calculated that $\tau_r \simeq 1/\nu_0$ even for the weak attractive forces of Van der Waals' binding. Sears and Cahn[47] considered the consequences of a large τ_r. However, in view of the work of Lennard-Jones and McFee the situation which they investigated does not seem important, at least for the case of high binding energy.

In connection with equation (A-8) one notes that the surface diffusion coefficient of the adsorbed molecules on the substrate is

$$D_s = a^2(kT/h)[1 - \exp(-h\nu/kT)] \exp(-\Delta G^\star_{\text{sd}}/kT), \tag{A-9}$$

* This consideration will be discussed in detail in Section D which deals with growth.

† The origin of this equation is discussed in Section D-1.

where a is the jump distance on the surface and $\Delta G^{\star}_{\mathrm{sd}}$ is the activational free energy for a surface-diffusion jump. According to the Einstein relation the adsorbed molecule will diffuse an r.m.s. distance before re-evaporating given by

$$\bar{X} = (2D_s\tau_s)^{\frac{1}{2}} = 2^{\frac{1}{2}}\, a \exp\,[(\Delta G^{\star}_{\mathrm{des}} - \Delta G^{\star}_{\mathrm{sd}})/2kT]. \qquad \text{(A-10)}$$

In summary, for metals incident on clean metallic substrates from monatomic vapors or in general for molecules incident on substrates with large attractive forces and no activational entropy for condensation, α_T should equal unity. In other cases α_T may differ from unity and may even have an apparent value greater than one.

(*c*) *Experiment*

Values of $\alpha_T < 1$ have been observed for various light gases on crystal substrates (see refs. 11, 13 and 48). However, when strong attractive forces are present, the evidence to date is that $\alpha_T = 1$.

In free evaporation, Stern,[(49)] with silver, and Rothberg *et al.*,[(50)] with single crystals of sodium chloride and lithium fluoride and polycrystalline cesium iodide and cesium bromide, found Maxwellian distributions of velocity corresponding to the source temperature, indicating $\alpha_T' = 1$.* Alty and Mackay[(51)] in an interesting experiment on evaporation of water found $\alpha_T' = 1$ even though the evaporation coefficient α_v was 0.036.

For impingement of potassium atoms upon magnesia and lithium fluoride substrates, Ellett and Cohen[(52)] found the reflected† beam to show a Maxwellian velocity distribution and $\alpha_T = 1$. Recently, McFee[(45)] demonstrated by very accurate experiments in high vacuum that $\alpha_T = 1$ and that a Maxwellian velocity distribution obtains in the reflected beam for the case of potassium incident on copper. Also he found a Maxwellian distribution and probably $\alpha_T = 1$ for potassium incident on tungsten, gold and magnesia. Further, he observed that the reflected beam of potassium from lithium fluoride was non-Maxwellian and that $\alpha_T \simeq 0.7$. However, there was evidence for chemical interaction of the potassium with the lithium fluoride substrate at high substrate temperatures and hence chemical reaction may have accounted for the value of $\alpha_T < 1$ in this case. McFee noted that his data for potassium on lithium fluoride are in disagreement with Ellett and Cohen's[(52)] and with Taylor's results[(22)] on the cosine law‡ and suggested that impurity adsorption may have led to complete thermal accommodation in their experiments.

Indirect evidence that $\alpha_T = 1$ is provided by experiments such as the emission-current measurements of Hughes[(53)] which show that $\tau_s \simeq 10^{-3}$ sec and hence is much greater than $1/\nu$ in the case of rubidium from an atomic

* Note that with an activation barrier for evaporation one would expect a Maxwellian velocity distribution but $\alpha_T' > 1$.

† As used here "reflected" does not necessarily imply elastic reflection.

‡ To be discussed in the following.

beam being adsorbed onto tungsten at 700–900°C. Similar work was carried out by Fraunfelder and co-workers[54] who used radioactive tracer techniques and high vacua of 10^{-9} to 10^{-10} mm Hg to determine binding energies and τ_s values for silver on molybdenum and nickel substrates. Their results show that the sticking coefficient of silver on both surfaces is temperature independent, which implies that $\alpha_c = \alpha_T = 1$. Also in this regard indirect evidence is provided by observations on the contamination of field-emitter tips (e.g. see ref. 15) which indicate that adsorption followed by surface diffusion and/or re-evaporation is the mechanism of interaction of the tips with residual gases. Finally, all cases cited in later sections wherein the condensation coefficient $\alpha_c = 1$ clearly require $\alpha_T = 1$.

In summary, evidence is compelling that $\alpha_T = 1$ for a molecule incident on a substrate in the case of strong binding forces, implying that the incident molecule adsorbs onto the substrate, thermally equilibrates there and leaves only by re-evaporation. Cases where one might expect a large activational entropy to lead to $\alpha_T < 1$ and reflection have not received extensive study. The role of impurity adsorption has yet to be clarified. In relation to the specific condensation coefficient, α_{c_1}, the prime conclusion is that an α_T of unity does not necessarily imply that $\alpha_{c_1} = 1$ but an α_{c_1} of unity requires that $\alpha_T = 1$. Further, an $\alpha_T = 0$ indicates that the incident molecules rebound elastically so that α_{c_1} must be zero also.

3. The Cosine Law

(a) Theory

The spatial distribution of molecules emitted (evaporated or reflected) from a substrate is another quantity related to the interaction of molecules and substrates. The reflection of a molecular beam is specular if the angle of incidence θ_I equals the angle of re-emission θ_R and the directions of incidence and reflection lie in a plane normal to the surface. The reflection is diffuse if the intensity of emitted molecules as a function of the angle of re-emission θ_R obeys the cosine law.[10] The cosine law can be developed from elementary kinetic theory of gases[1] and is the equivalent of Lambert's law in optics.* Diffuse emission can result from thermal equilibration of the incident molecules with the surface, complete momentum accommodation, or scattering due to surface asperities. In complete momentum accommodation the horizontal components of momentum of a reflected or re-evaporated

* The cosine law states that the vaporizing flux $J(r, \theta_R)$ at a distance r from a differential source of vaporization is directly proportional to $\cos \theta_R$

$$J(r, \theta_R) = J_v(dA/\pi r^2) \cos \theta_R$$

where dA is the element of area at the source and θ_R is the angle between the direction of molecules from the source and the normal to the surface.

Violation of the cosine law is possible only in a directed-beam experiment. In the case of a substrate exposed to an equilibrated vapor phase, the cosine law will always obtain for the total flux emitted from the surface, i.e. the sum of reflection and re-evaporation fluxes.

molecule are unrelated to those of the incident molecule. For a specularly reflected molecule, the components of momentum parallel to the surface are the same for the molecule during incidence and after reflection.* For scatter from surface asperities the condition

$$h_I \cos \theta_I > \lambda_I \tag{A-11}$$

must be satisfied in which h_I is the step height of the asperities and λ_I is the de Broglie wave length of the incident molecular beam.

The de Broglie wave length for hydrogen at room temperature is about one angstrom and for any material varies as the reciprocal square root of the product of molecular mass and temperature. Mechanically polished surfaces have irregularities of the order 10^{-5} cm, and cleaved surfaces have irregularities of multiple or mono-molecular step heights of 2 to 20 × 10^{-8} cm. To date only whiskers and platelets (e.g. ref. 16) appear to have perfect low-index surfaces, and even such perfect surfaces are "roughened" by vibrations of surface atoms. Hence in view of equation (A-11) one expects diffuse reflection for all but the lightest gases at beam temperatures of 300°K or above and even for those gases when $\theta_I \ll \pi/2$.

Gaede[17] and Knudsen[10,18] pointed out that the distribution of energies and directions of molecules emerging from a stationary surface and entering an equilibrated vapor phase cannot differ from the distribution of incident molecules without a violation of the second law of thermodynamics. Millikan,[19] following Gaede, noted that since the distribution of impinging molecules was Maxwellian, the cosine law of reflection should follow from the second law criterion. His example is depicted in Fig. 3. If surface B is rough, giving diffuse or cosine reflection, and surface A is smooth and reflects molecules in, say, a direction normal to the surface, a net moment will result. This gives rise to rotation about 0 and hence a violation of the second law. Epstein[20] followed Millikan's suggestion and demonstrated that the cosine law of reflection must obtain in the case of a surface equilibrated with the vapor phase and that cosine law reflection could result from (a) specular reflection, (b) diffuse reflection, or (c) diffuse reflection with accommodation. It is now appreciated that case (c) is indistinguishable from adsorption and re-evaporation.

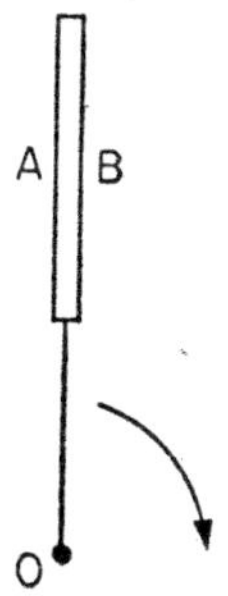

FIG. 3. Millikan's thought experiment to prove the cosine law.

Knudsen[18] postulated that the cosine law obtains for free evaporation as well. In the absence of any surface constraints, it may be shown from an extension of Epstein's arguments and the principle of microscopic reversibility[5] that the cosine law is to be expected in free evaporation. However, when surface constraints introduce entropies of activation into the evaporation process, Epstein's arguments can no longer be applied and deviations from the cosine law are possible. For example, reconsider Fig. 1. If the

* Note that complete momentum accommodation does not necessitate thermal equilibration ($\alpha_T = 1$).

surface is highly polarized, a polar molecule might arrive from the equilibrium vapor in a manner such that it senses either potential curve (1) or (2), depending on its orientation. Thus it may reflect in case (2) and adsorb in case (1), and only the component corresponding to case (1) will be detected in free evaporation. Due to the directional sensitivity of the potential energy curve in such a case it is conceivable that this component could show deviation from the cosine law.

In summary, the cosine law must obtain for a surface equilibrated with the vapor phase and thus violation of the cosine law is possible only in a directed-beam experiment. The cosine law should hold for a directed molecular beam because of scatter from surface asperities except for the lightest gases at high angles of incidence and low temperatures. This conclusion is, of course, independent of whether the molecules are reflected or adsorbed and re-evaporated. It is apparent from the above that compliance with the cosine law is not a sensitive criterion for either energy or momentum accommodation of an impinging molecular beam. Also it is evident that adherence to the cosine law is no indication that the specific condensation coefficient α_{c_1} equals unity. However, specular reflection suggests that $\alpha_{T'} < 1$ and proves that $\alpha_{c_1} < 1$. Finally, the cosine law is probably valid for free evaporation of simple solids or liquids which have a monatomic vapor phase and hence are unlikely to evaporate with appreciable entropies of activation. However, non-compliance with the cosine law in free evaporation suggests that both $\alpha_{T'}$ and α_{c_1} are less than unity.

(*b*) *Experiment*

Using the apparatus represented in Fig. 4 Knudsen,[10] in the case of mercury, and Wood,[21] for mercury and cadmium, showed that the cosine law holds for re-emission of these metal atoms after impingement on glass.

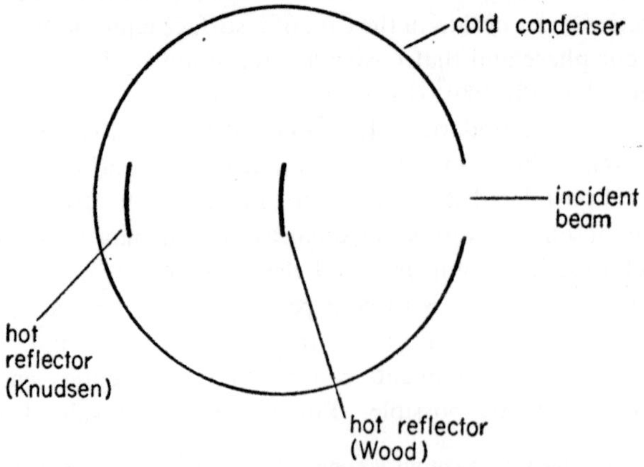

FIG. 4. Knudsen's and Wood's experiments to prove cosine law.

Taylor[22] later demonstrated that the cosine law obtains for beams of potassium, lithium and cesium incident on lithium fluoride and sodium chloride. In the experiments of Knudsen[10] and Wood,[21] although this is the classical work in the field, surface roughness could have led to the cosine law. Also both the glass reflector and the cold condenser were probably contaminated by adsorbed impurity gas. Even in Taylor's[22] work, where contamination was less likely, some surface scattering was possible due to surface roughness. Bassett *et al.*,[23] investigating surface smoothness of alkali halides, have shown that even cleavage surfaces have roughness due to the presence of mono-molecular ledges over portions of the surface.

On the other hand helium on lithium fluoride substrates[24,25] and hydrogen and helium on sodium chloride substrates[26] show marked deviations from the cosine law in the direction of specular reflection. As predicted by equation (A-11) the specular reflection was more pronounced at high angles of incidence and at low temperatures. Moreover, Josephy[27] found appreciable specular reflection of mercury incident on cleavage surfaces of lithium fluoride and sodium chloride under conditions where, for the mercury, $\lambda_I \simeq 0.1$ Å and hence surface scattering might have been expected. Similar results have been obtained[28-32] for reflection of zinc, cadmium, arsenic, lead, antimony and thallium from alkali halide crystals. In Hancox' work[28] he found that it was necessary to outgas cleaved lithium fluoride at 450°C (after exposure to air) in order to obtain specular reflection, indicating that adsorbed films affect the fulfillment of the cosine law. Although the surface topographies of the substrates were not examined in these experiments, it must be supposed that appreciable areas of the surfaces were free from the mono-molecular ledges observed by Bassett *et al.*[23]

Hurlbut[33] has tested the cosine law for beams of air, nitrogen and argon incident on teflon, glass, aluminum and steel. The importance of the substrate was noted in nitrogen-beam experiments where the cosine law was found to be valid for aluminum and steel substrates but not for glass. Hurlbut also found partial specular reflection of air, nitrogen and argon from teflon and optically polished glass and observed that the amount of specular reflection increased with angle of incidence. Hurlbut's experiments were carried out in high vacua and with outgassed substrates.

Thus far we have been discussing the reflection of thermal molecular beams. That the energy of the incident particles may have an effect in the higher portion of the energy spectrum is indicated by the results of sputtering experiments[34] which show that high-energy incident ions are reflected in directions parallel to low-index directions of the substrate. However, the energies in these experiments were much higher than thermal energies (150 eV versus ~0.01 eV for thermal beams). Specular reflection of neutral molecules in the high-energy portion of a thermal spectrum has not been studied.

Finally, Knudsen[18] has indicated that the cosine law holds in free evaporation of sulfur, zinc and silver. However, he found some indication that deviation occurred in the case of sulfur, and his experimental technique was only sensitive to $\sim \pm 50\%$ in beam intensity as a function of θ_R. There have been no experiments on free evaporation in high vacua of oriented single crystals of materials with large activational entropies of evaporation, conditions which would favor deviation from the cosine law.

In summary, evidence for specular reflection from alkali halide cleavage surfaces and from polished teflon and glass exists, but agreement between different authors or between theory and experiment is poor. It is not possible at present to discern whether partial thermal accommodation accompanies partial specular reflection although such a circumstance is possible. Also, no experiment has been done in which specular reflection and a non-Maxwellian velocity distribution have been found concurrently. The observed diffuse reflection from other surfaces may be due to thermal accommodation, momentum accommodation in the absence of thermal accommodation, or scatter from surface asperities. Even the evidence for the validity of the concept of scatter from surface asperities is contradictory. Evidence exists that surface roughness, substrate crystal structure, substrate temperature, cleanliness, beam angle of incidence and beam temperature all affect the degree to which the cosine law obtains. Experiments on the cosine law in free evaporation are sparse and inconclusive.

Largely on the basis of theory it is concluded that compliance with the cosine law is not a sensitive criterion for either thermal or momentum accommodation of an impinging molecular beam. Also it is evident that adherence to the cosine law does not indicate that the specific condensation coefficient α_{c_1} is unity. On the other hand specular reflection suggests that the thermal accommodation coefficient α_T is less than unity and proves that $\alpha_{c_1} < 1$. Finally non-compliance with the cosine law in free evaporation suggests that both α_T and α_{c_1} are less than one.

4. Summary of Relationships between Coefficients of Condensation, Evaporation and Thermal Accommodation and Other Factors

A summary of the various possible interactions between incident molecules and substrates is presented in Table 1. There are forty-five different possible combinations of the five varying factors: specific condensation coefficient α_{c_1}, thermal accommodation coefficient α_T, the extent X of compliance with the cosine law for diffuse reflection, the degree M of momentum accommodation, and the extent A to which the cosine law would obtain due to scattering from surface asperities alone. Thus, for example, Case 2-c-(1) represents partial thermal accommodation. α_{c_1} may be zero if the incident molecule rebounds into the vapor but gives up some energy inelastically during collision; or α_{c_1} may be less than unity but finite if some molecules rebound into the vapor but others are adsorbed onto the surface. The cosine law is

TABLE 1

1. $\alpha_T = 0$, $\alpha_{c_1} = 0$.
   a. $X = 0$, $M = 0$, $A = 0$.
   b. $X = f$.
      (1) $M = f$, $A = 0$ or f.
      (2) $A = f$, $M = 0$ or f.
   c. $X = 1$.
      (1) $M = 1$, $A = 0, f$ or 1.
      (2) $A = 1$, $M = 0, f$ or 1.

2. $\alpha_T = f$, $\alpha_{c_1} = 0$ or f.
   a. $X = 0$, $M = 0$, $A = 0$.
   b. $X = f$.
      (1) $M = f$, $A = 0$ or f.
      (2) $A = f$, $M = 0$ or f.
   c. $X = 1$.
      (1) $M = 1$, $A = 0, f$ or 1.
      (2) $A = 1$, $M = 0, f$ or 1.

3. $\alpha_T = 1$, $\alpha_{c_1} = 0, f$ or 1.
   a. $X = 0$, $M = 0$, $A = 0$.
   b. $X = f$.
      (1) $M = f$, $A = 0$ or f.
      (2) $A = f$, $M = 0$ or f.
   c. $X = 1$.
      (1) $M = 1$, $A = 0, f$ or 1.
      (2) $A = 1$, $M = 0, f$ or 1.

Tabulation of possible combinations of various factors dependent on the interaction with a substrate of incident molecules from the vapor. The degree to which a given property exists is represented by 1, f or 0 for complete, partial or zero fulfillment. α_T represents the thermal accommodation coefficient, α_{c_1} the specific condensation coefficient for the constraint of elastic reflection, X the extent of compliance with the cosine law, M the degree of momentum accommodation, and A the extent to which the cosine law would obtain due to scattering from surface asperities alone.

obeyed ($X = 1$) because of complete momentum accommodation ($M = 1$) independent of the degree of surface roughening ($A = 0, f$ or 1). Some of the cases are possible but unlikely, such as the case in which $\alpha_T = 1$ and $\alpha_{c_1} = 0$. This could occur if the binding energy of a molecule to a substrate were small compared to kT. Then even though an incident molecule equilibrated thermally with the substrate it would re-evaporate on its first surface vibration, thus apparently reflecting.

In conclusion the following relations exist between the various factors of interest in this discussion.

(i) $\alpha_c = \prod_i \alpha_{c_i}$ and $\alpha_v = \prod_i \alpha_{v_i}$ are, in general, not equal. However, for some specific constraints α_{c_i} may equal α_{v_i}, such as for the constraint of elastic reflection where $\alpha_{c_1} = \alpha_{v_1}$. It is not yet possible to give a theoretical equation for α_{c_1}. However, quantitative expressions for other α_{c_i}'s and α_{v_i}'s will be developed in Sections D and E which deal with the processes of growth and evaporation.

(ii) Evidence is compelling that $\alpha_T = 1$ for a molecule incident on a substrate in the case of strong binding forces, implying that the incident molecule adsorbs onto the substrate, thermally equilibrates there and leaves only by re-evaporation, i.e. that $\alpha_{c_1} = 1$.

(iii) An α_T of unity does not necessarily imply that $\alpha_{c_1} = 1$ but an α_{c_1} of unity requires that $\alpha_T = 1$. Further, an $\alpha_T = 0$ indicates that the incident molecules rebound elastically so that α_{c_1} must be zero also.

(iv) Compliance with the cosine law for diffuse scattering is not a sensitive criterion for either thermal or momentum accommodation of an impinging molecular beam. Also, adherence to the cosine law does not indicate that α_{c_1} is unity. However, specular reflection suggests that α_T is less than one and proves that $\alpha_{c_1} < 1$. Finally, non compliance with the cosine law in free evaporation suggests that both α_T and α_{c_1} are less than one.

B. NUCLEATION IN THE VAPOR PHASE

1. Introduction

A NUMBER of reviews have appeared on the general subject of nucleation in phase transitions.[55–64] Some of these, in particular the articles by Turnbull and Hollomon,[56] Bradley,[57] LaMer,[58] Pound,[59] Hollomon and Turnbull,[60] Turnbull[62] and Courtney[64] have dealt with nucleation in the vapor phase. However, recent experimental and theoretical developments, in particular the work of Lothe and Pound,[65] indicate that large corrections must be made to previous treatments of nucleation theory. Also, most of the above review papers have dealt with homogeneous* nucleation in the vapor, a well justified position in view of the extreme relevance of this case to the establishment of nucleation theory in general. Nevertheless, it now appears that almost all of the available critical supersaturation data represent heterogeneous nucleation on stable molecular aggregates about gaseous ions. Accordingly it would seem that a general discussion of heterogeneous nucleation in the vapor is in order.

Therefore, Parts 2 and 3 of this section will deal with the theory of homogeneous and heterogeneous nucleation in the vapor, respectively, and Part 4 will discuss the relevant experimental findings. It will emerge that only in the case of homogeneous nucleation is the theory simple enough for a critical comparison with experiment while, unfortunately, most of the available data relate to heterogeneous nucleation as noted above. Further, all of the present nucleation theory which is in a form suitable for quantitative comparison with experimental data is encumbered by the gross assumption inherent in assigning macroscopic thermodynamic quantities to clusters involving only a few molecules.

2. Theory of Homogeneous Nucleation

(a) *Theory of Homogeneous Nucleation of Droplets from Unary Vapors*

1. Steady-State Theories

(*a*) *Volmer–Weber–Becker–Doering theory.* Steady-state theories of nucleation consider the growth of clusters from a supersaturated vapor of

* Homogeneous nucleation in the vapor occurs without the aid of catalysis by gaseous ions, gaseous reactive chemical impurities or suspended solid or liquid particles which are said to produce heterogeneous nucleation.

single molecules by a series of bimolecular reactions in which the clusters grow by addition of one molecule per reaction.* They develop initially with an increase in free energy until a critical size is reached, above which growth continues with a decrease in free energy. The nucleation rate is then considered to be the product of the concentration of critical nuclei and the frequency with which they grow by addition of one molecule. Steady state is achieved by the device of having a Maxwell demon take clusters which have grown larger than the critical and sunder them into single molecules. The consequences and applicability of this model will be examined in the following. First, attention is focussed on the free energy of formation of the critical cluster.

Subject to the macroscopic assumption noted above about the properties of clusters, the surface and volume contributions to the Gibbs free energy of formation of a spherical cluster of molecules, which is envisaged as an embryo of the stable liquid phase, may be represented by

$$\Delta G^0 = 4\pi r^2 \sigma + 4\pi r^3 \Delta G_v / 3 \tag{B-1}$$

where r is embryo radius, σ is the specific interfacial free energy and

$$\Delta G_v = -(kT/\Omega) \ln (p/p_e) \tag{B-2}$$

is the Gibbs free energy difference per unit volume of liquid between the supersaturated vapor of pressure p and bulk liquid of equilibrium vapor

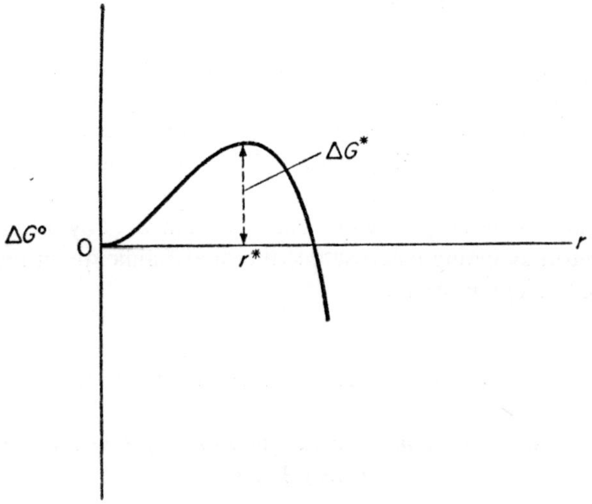

FIG. 5. Plot of equation (B-1).

pressure p_e and molecular volume Ω. It is seen from the plot of equation (B-1) shown in Fig. 5 that this free energy of formation increases with r to

* As noted by Volmer and Weber[66] and Kaischew and Stranski,[67] collisions between clusters are highly improbable and thus clusters of molecules (droplets) must form in the vapor by successive additions of monomeric molecules.

approach a maximum $\Delta G^\star$ at $r^\star$, the radius of the critical nucleus. One speaks of clusters smaller than $r^\star$ as unstable embryos while larger aggregates are termed stable nuclei of the liquid phase. Upon maximizing (B-1) with respect to r one gets

$$r^\star = -2\sigma/\Delta G_v, \tag{B-3}$$

the Gibbs–Thomson relationship, and

$$\Delta G^\star = 16\pi\sigma^3/3\Delta G_v^2 \tag{B-4}$$

which was first derived by Gibbs[68] who carefully restricted its application to large droplets only. The number of molecules in a cluster is related to the radius r by

$$i = (4\pi r^3/3\Omega). \tag{B-5}$$

Thus the condition for the critical nucleus is

$$(\partial\Delta G^0/\partial r) = (\partial\Delta G^0/\partial i) = (\mu_c - \mu_v) = 0 \tag{B-6}$$

where μ_c and μ_v are the chemical potentials of molecules in the nucleus and vapor, respectively. Therefore the chemical potentials of molecules in the critical nucleus and in the supersaturated vapor are equal although formation of a nucleus occurs with local increase in free energy as shown in Fig. 5. This latter circumstance is possible only through a statistical fluctuation involving a limited number of molecules.

The concentration of single molecules will be much greater than the sum of the concentrations of all other size classes* so that the equilibrium constant for a cluster containing i molecules is

$$[n_i/(n_1 + \sum_2^i n_i)]/[n_1/(n_1 + \sum_2^i n_i)]^i = (n_i/n_1) \tag{B-7}$$

where n_i is the concentration of clusters containing i molecules and $n_1 = (p/kT)$ is the concentration of monomer. The metastable equilibrium concentration of critical nuclei in terms of the van't Hoff reaction isotherm is then[66]

$$n_i^* = n_1 \exp(-\Delta G^\star/kT). \tag{B-8}$$

On the basis of the usual assumption in physico-chemical kinetics that the actual concentration of activated species is the metastable equilibrium one, Volmer and Weber[66] formulated the nucleation kinetics in terms of the molecular impingement rate on this equilibrium concentration of critical nuclei. Their equation for homogeneous nucleation rate is

$$J = A^\star \omega n_i^\star \tag{B-9}$$

* This condition and equation (B-7) are very good approximations for nucleation from the vapor. In nucleation in condensed systems, $\sum_2^i n_i$ may not be negligible compared to n_1, as evidenced in the work of Sundquist and Oriani.[69]

where $A^\star = 4\pi r^{\star 2}$ and the frequency factor for impingement

$$\omega = \alpha_c p/(2\pi mkT)^{\frac{1}{2}}$$

giving

$$J = \alpha_c(4\pi r^{\star 2})[p/(2\pi mkT)^{\frac{1}{2}}]n_1 \exp(-\Delta G^\star/kT). \tag{B-10}$$

Becker and Doering[70] and Zeldovich[71–73] refined this treatment by calculating the actual steady-state concentration and gradient of critical nuclei in which they evaluated loss of critical nuclei both by promotion to supercritical size (growth to become macroscopic droplets) and decomposition by loss of molecules (to become subcritical embryos). Also they suggested that the condensation coefficient α_c for the molecules impingent on the clusters might be appreciably less than unity. In their treatments, rather than assuming equilibrium which leads to equation (B-8), they obtained a solution for a set of bimolecular reactions of the type

$$B_{i-1} + B \rightleftharpoons B_i \tag{B-11}$$

where B_i represents the ith-sized cluster. The solution to this set of equations[70–73] yields the so-called non-equilibrium factor, which is given by

$$Z = (\Delta G^\star/3\pi kTi^{\star 2})^{\frac{1}{2}} = \Omega\Delta G_v^2/8\pi\sigma(\sigma kT)^{\frac{1}{2}} \tag{B-12}$$

in which $i^\star$ is the number of single molecules in the critical nucleus.* Under the usual experimental conditions Z has values of the order of 10^{-2}. Accordingly the Becker–Doering expression for rate of homogeneous nucleation of droplets from the vapor becomes

$$J = \alpha_c Z(4\pi r^{\star 2})[p/(2\pi mkT)^{\frac{1}{2}}]n_1 \exp(-\Delta G^\star/kT). \tag{B-13}$$

It is interesting to note that equation (B-13) is one of the few physico-chemical rate expressions which explicitly takes into account reduction of activated species below the equilibrium concentration, and this in terms of some 100 consecutive reactions.

Volmer[4] inserted another term, $\exp(\Delta H_{\text{vap}}/kT)$ which involves the molecular enthalpy of vaporization, into the rate expression. This term has values of the order of 10^6. However, in view of the more elegant Zeldovich[70–73] derivation of equation (B-13), it appears that there is no basis for including such a term. Further, Volmer evaluated α_c by considering the requirement of dissipation of energy of condensation. He reasoned that unless this energy were removed from the smaller embryos within the period of molecular vibration, decomposition of the embryos would occur. According to this very approximate treatment, $\alpha_c \simeq 10^{-6}$, and therefore he cancelled α_c with $\exp(\Delta H_{\text{vap}}/kT)$ and omitted both terms from equation (B-13).

* A somewhat more complicated expression for Z has been given by Barnard.[74] Also it is noted that equation (B-12) has been stated incorrectly in a number of standard nucleation references.

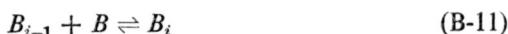

The present authors feel that this is unjustified and think that exp $(\Delta H_{vap}/kT)$ should be omitted and α_c included in equation (B-13). In fact, as will be discussed in a later section, it appears that theoretical evaluation of α_c deserves more careful attention.

(*b*) *Recent modifications of the steady-state theory.* There is a fundamental difficulty with equation (B-13) which has received extensive study; it contains the free energy of formation of critical nuclei as given by equation (B-4) in terms of macroscopic thermodynamic quantities such as liquid density, surface tension and vapor pressure, whereas in actuality the critical nuclei are comprised of only a few molecules. In fact according to equation (B-3) $r^\star \simeq 10$ Å, and the number of molecules in the critical nucleus is of the order of 50 in most experimental situations. This problem has been treated by a number of authors[75–83] who have developed statistical mechanical treatments of the thermodynamics of small clusters. One important problem is in the application of standard capillarity theory to free surfaces of high curvature and the attendant difficulty in locating the surface of tension. The classical theory outlined above assumes that the surface of tension is identical with that of density discontinuity at r. Using a quasi-thermodynamic approach, Tolman[76] concluded that the surface of tension is located beneath the surface of density discontinuity in small droplets and predicted a decrease in surface tension at high curvatures. These results were supported by a statistical mechanical treatment of surfaces of low curvature due to Kirkwood and Buff.[77] Their expression is

$$\sigma = \sigma_0/[1 + (2\delta'/r)] \tag{B-14}$$

where σ_0 is the surface tension at zero curvature and δ' is the displacement of the surface of tension beneath the surface of discontinuity and is approximately of molecular dimensions. If equation (B-14) is applied to droplets of the size of critical nuclei, a reduction in surface tension of approximately 20% is predicted. However, Buff[78] has shown that equation (B-14) is not applicable for such high curvatures and in fact that present theory does not cope with the problem at hand.

Reiss[79] attempted to circumvent the ambiguities attending assumption of an abrupt surface of discontinuity and application of macroscopic thermodynamic concepts to nucleation kinetics by treating the clustering problem in terms of statistical mechanics. According to his model the vapor system was divided into spatial cells, each containing the number of molecules necessary to form a critical nucleus, and the probability of nucleus formation was calculated. However, this model did not permit of transport of vapor molecules between cells, and hence the predicted nucleation rates were too low. Further, mathematical complexity has precluded efforts to remove the restriction of isolated cells.[80]

Another new and promising avenue of approach to the problem of describing the activated species in nucleation is through the thermodynamics of inhomogeneous systems. The fundamentals of this subject have been developed by Cahn and Hilliard,[81] Hart,[82] and Cahn,[83] who derived an expression for the free energy of a non-uniform system and applied it to calculate the interfacial tension between two coexisting phases. In this method the free energy is given as a function of the derivatives of density or composition. Their equation for the tension at a planar interface between two phases at equilibrium is

$$\sigma = \int (\Delta f' + Kg^2)\, \mathrm{d}x \tag{B-15}$$

where $\Delta f'$ is the free energy of material of the local density or composition in the absence of a gradient, g is the gradient of density or composition in the x direction normal to the interface, and K is a constant which is determined principally by the intermolecular forces in the particular system. The general expression for Gibbs free energy of formation of a critical nucleus in fluid systems is

$$\Delta G^{\star} = \int (\Delta f' + Kg^2)\, \mathrm{d}V. \tag{B-16}$$

This integral has been evaluated for nucleus formation in both unary and binary incompressible fluid systems,[84] but not for the present case in which a compressible fluid is involved. In any case, comparison of this theory with experiment would require detailed knowledge of intermolecular forces. Finally one notes that there is some question in regard to a major assumption of this theory, i.e. that the gradients of density or composition are not large. It would seem that this condition is unlikely at nucleus–vapor interfaces under most experimental conditions.

The so-called Becker–Doering equation (B-13) has been applied almost universally to analysis of droplet nucleation observations, and for many years its agreement with the critical supersaturation data of Volmer and Flood[85] stood as the foundation of our knowledge of nucleation. However, within the past year Lothe and Pound[65] showed that there are highly important statistical mechanical contributions to the free energy of formation of embryos and critical nuclei which had been previously neglected.* The least important of these is the free energy of separating a group of i molecules from a larger ensemble and is given approximately by

$$\Delta G_s \simeq kT \ln (2\pi i)/2 + Ts \tag{B-17}$$

where s is the molecular entropy of the liquid, excluding contributions from the energy. Two much more important contributions arise from consideration of the absolute entropy of the embryos or nuclei in the vapor phase. $\Delta G^{\star}$ of equation (B-4) is the free energy of formation of a critical nucleus

* Courtney[86] showed that the classical equation (9) neglected the free energy change required to compress nuclei from their standard state pressure p_e to p. This small correction term is included in the Lothe–Pound treatment.

at rest. The true free energy of formation must include the translational and rotational components necessary to energize the nuclei to temperature T in the standard-state vapor of embryos at p and T. The translational component for an embryo or nucleus is given by

$$\Delta G_t = -kT \ln [(2\pi m_i kT)^{\frac{3}{2}} \Omega_i / h^3] \qquad \text{(B-18)}$$

in which m_i is molecular mass and $\Omega_i = (kT/p)$ is the molecular volume of the embryo in the vapor at p and T. The rotational component is

$$\Delta G_r = -kT \ln [(2kT)^{\frac{3}{2}} (\pi I^3)^{\frac{1}{2}} / \hbar^3] \qquad \text{(B-19)}$$

where I is the moment of inertia of the spherical droplet and $\hbar = h/2\pi$. A similar term relating to the absolute rotational entropy of molecules in a liquid must be considered in homogeneous nucleation of crystallites from liquids. The terms discussed above, equations (B-1), (B-17), (B-18) and (B-19) may be summed to give the Gibbs free energy of formation of an embryo in the vapor*

$$\Delta G' = \Delta G^0 + \Delta G_s + \Delta G_t + \Delta G_r. \qquad \text{(B-20)}$$

This expression may be maximized with respect to embryo size to yield $r^\star$ and $\Delta G^{\star\prime}$, the radius and free energy of formation of the critical nucleus. However, the last three terms are insensitive to size,† so it suffices to take

$$\Delta G^{\star\prime} = \Delta G^\star + \Delta G_s + \Delta G_t + \Delta G_r \qquad \text{(B-21)}$$

which should replace $\Delta G^\star$ in equation (B-13). Further, since only the first term varies markedly with supersaturation, it is convenient to bring the last three terms into the pre-exponential and hence equation (B-13) becomes

$$J = \alpha_c Z' \Gamma (4\pi r^{\star 2}) [p/(2\pi mkT)^{\frac{1}{2}}] n_1 \exp(-\Delta G^\star/kT) \qquad \text{(B-22)}$$

where Γ is the Lothe–Pound correction factor

$$\Gamma = (IkT)^{\frac{3}{2}} (kT/p)(i^\star/\pi^2)(2\pi mkT)^{\frac{3}{2}} \exp(-s/k)/\hbar^6. \qquad \text{(B-23)}$$

In this situation the non-equilibrium factor $Z' \simeq (\Delta G^{\star\prime}/3\pi kTi^{\star 2})^{\frac{1}{2}}$, cf. equation (B-12), again typically has values of the order of 10^{-2}. In the case of homogeneous(88–91) nucleation of droplets from water vapor at 300°K, the number of molecules in the critical nucleus $i^\star \simeq 100$, $m = 3.0 \times 10^{-21}$ g, $I = 8.6 \times 10^{-36}$ g-cm^2, $p = 0.075$ atmospheres and $\Omega_i = 5.5 \times 10^{-19}$ cm^3. Thus in this case $\Delta G_s \simeq 8kT$, $\Delta G_t = -24.4\, kT$ and $\Delta G_r = -20.8\, kT$, resulting in a total quantum statistical contribution to $\Delta G^{\star\prime}$ of $-37.2\, kT$.

* It is conceivable that relaxed surface vibrations could also affect $\Delta G'$.(87) These terms would appear in the surface free energy and hence would affect the degree to which the assumption holds that the macroscopic value may be assigned to the surface of an embryo.

† In other words the Gibbs–Thomson relationship (B-3) is still valid to an excellent approximation.

This corresponds to an increase in metastable equilibrium concentration of critical nuclei and hence nucleation rate by a factor of $\Gamma = 10^{17}$.

2. *Non-Steady-State Theories*

As will be discussed in the section on experimental results, it is obvious that the recent theoretical development described above has destroyed agreement between theory as expressed by equation (B-22) and critical supersaturation measurements from cloud-chamber work. Indeed, the theoretical estimate of critical supersaturation ratio now appears to be too low by a factor of 1.3. At first sight one might be inclined to blame most of this large discrepancy on the inadequacy of equation (B-4) for $\Delta G^\star$ to describe the critical nucleus because of the assumption of macroscopic thermodynamic properties for microscopic clusters which is inherent in its derivation. However, as shall be shown in a later chapter, the classical theory describes the homogeneous nucleation of bubbles in ebullition quite well. This suggests that much of the disagreement in the case of nucleation of droplets from vapor may be due to thermal non-accommodation in the clustering process, which would result in $\alpha_c < 1$. In fact it seems possible, particularly in view of the brief periods (of the order of 1/10 sec) of supersaturation in most cloud chambers, that a transient rather than a steady-state problem may be at hand. In relation to this suggestion, Zeldovich,(92) Kantrowitz,(93) Probstein,(94) Collins,(95) Wakeshima,(96) Turnbull,(97) Frisch(98) and Courtney(64) have considered the time period required to build up a steady-state population of embryos upon imposing a given supersaturation. All neglected the contributions to the free energy of formation of an embryo introduced by Lothe and Pound. The first five authors approximated the set of equations (B-11) by a differential equation for the time dependence of n_i, and all are in essential agreement that the induction period to establish the steady state is

$$\tau_K \cong 2\pi kT/A^\star\omega(\partial^2\Delta G'/\partial i^2)_{i=i^\star} \cong i^{\star 2}/[\alpha_c(4\pi r^{\star 2})p/(2\pi mkT)^{\frac{1}{2}}]. \qquad \text{(B-24)}$$

In the usual experimental situations, and with the condensation coefficient $\alpha_c \cong 1$, τ_K assumes values of the order of 10^{-7} sec. Thus it might appear that the period to achieve a steady-state population of embryos is insignificant.

Frisch(98) used a partial-difference differential equation to describe the time dependence of n_i and obtained a result equivalent to equation (B-24) for the case in which clusters larger than the monomer exist at time zero. However, Frisch calculated that there should be a large time lag if only monomer is present initially. Courtney(64) has criticized Frisch's treatment and repeated the calculations with the result that τ_K again approaches the value given by equation (B-24) if $\alpha_c \cong 1$. At any rate, as noted by Courtney,(64) the results of Fisher, Holloman and Turnbull(99) predict that some clusters will always pre-exist in the saturated vapor and hence equation (B-24) should hold even according to Frisch.(98)

Courtney(64) has solved for the time dependence of equations (B-11) by a computer solution of 100 differential equations for the time dependence of concentration of clusters of size $i = 1$ to $i = 100$. His results for the case of water are shown in Fig. 6. Again it is assumed that $\alpha_c \simeq 1$ for a monomer colliding with a cluster. It is seen that his results are qualitatively in agreement with equation (B-13). Courtney's work, which is continuing, appears quite promising.

However, all of the above workers approached the transient problem in terms of a collision picture and ignored possible thermal non-accommodation,

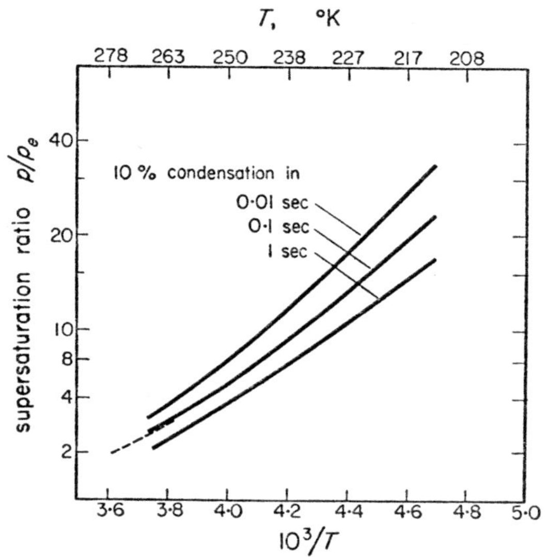

FIG. 6. Results of Courtney[64] compared with the prediction of the Becker–Doering equation (B-13) for condensation of water. - - - equation (B-13), ——— Courtney's calculated results.

which could affect equation (B-24) by reducing α_c (Chapter A). Furthermore thermal non-accommodation could be more important for small clusters than for large ones, as discussed by Lothe and Pound.[65] In such a case α_c would be dependent on size and the solution for the kinetics of cluster formation would be very formidable indeed without independent knowledge of α_c as a function of i. Perhaps the transient problem should be reviewed with consideration for the simultaneous and interdependent processes of diffusion of heat of vaporization (accommodation) and diffusion of clusters between the size classes. The present authors can only conclude at present that relationships of the type of equation (B-24), with the proper impingement frequency replacing $p/\sqrt{2\pi mkT}$, are valid only for time lag in nucleation in condensed phases where thermal accommodation is likely. If and when the problem of the induction period is solved, equation (B-22) would be modified by an additional factor $\Psi(t)$, where $\Psi(t) = 1$ for $t > \tau_K$, $\Psi(t) < 1$ for $t < \tau_K$.

(b) *Summary of Theory of Homogeneous Nucleation of Droplets from Unary Vapors*

A recent theoretical development,[65] involving statistical mechanical contributions to the free energy of formation of embryos, has negated the long-standing agreement between theory and critical supersaturation measurements from cloud-chamber work. Now equation (B-22) replaces the original Becker–Doering relationship (B-13). The discrepancy between theory and experiment may be attributed to the problem of non-accommodation in small clusters and to the cardinal assumption involved in assigning macroscopic thermodynamic quantities to microscopic systems. If successful, a treatment of non-accommodation might yield a size-dependent condensation coefficient α_c sufficiently less than unity to bring about agreement with observation.

(c) *Theory of Homogeneous Nucleation of Crystals from Unary Vapors*

In droplet nucleation, due to the isotropic surface tension, one could assume according to the classical theory that the embryos and nuclei were spherical and that their chemical potential was completely defined by a single continuous parameter, either the radius of curvature or the number of molecules in the embryo. Also, due to the high mobility of molecules in the liquid, the only molecular process to be considered was absorption at the cluster surface.

However, in crystallite nucleation the specific surface free energy σ is generally anisotropic. Thus the clusters should not be spherical and, in the case of large crystallites at least, the metastable equilibrium shape* is described by the well-known Gibbs–Wulff theorem.[100–102] Also the free energy of the crystallite due to the presence of the surfaces is given by

$$G_A = \int \sigma \, dA. \tag{B-25}$$

However, for small crystalline nuclei the contribution of kinks, steps, edges and corners should be appreciable. Further, as first noted by Gibbs, the chemical potential of a small faceted crystallite will not be a continuous function of the number of molecules it contains; for example, completion of a line or plane of atoms will cause an abrupt decrease in chemical potential while initiation of a line or plane will cause an abrupt increase in chemical potential. Also, in crystallite nucleation the mobility of molecules *in* the solid is relatively low and in general the mechanism of addition of molecules to a faceted crystallite must be adsorption, nucleation of two-dimensional discs, surface diffusion and accretion at steps as described in Section D-3.

* For moderate supersaturations.

Accordingly at least four molecular processes must be considered in crystal nucleation from the vapor in cases where the embryos and critical nucleus are faceted.

This more difficult case has been treated in an approximate manner by Stranski and Kaischew[103,104] who considered two-dimensional disc nucleation to be the only slow molecular process on the surface of a cubical embryo. Their result is analogous to the approximate equation (B-10) for droplet nucleation. A more complete kinetic derivation has been accomplished by Becker and Doering[4,70] who assumed a cubic homopolar lattice and represented the free energy of the embryo in terms of nearest-neighbor interaction,* thus dispensing with the necessity for determining cluster shape and free energy in terms of the Gibbs–Wulff theorem. In fact they assumed neither restriction on the mode of addition of molecules nor any particular process of rearrangement into optimum shapes by surface migration but considered all conceivable shapes and transitions between them. They were able to solve this very complex problem in terms of relationships developed for the electrical circuit theory of wire nets. Their resulting rate equation includes a non-equilibrium factor and is quite analogous to equation (B-13) for droplet nucleation.

Both treatments have inherent in them all the difficulties involved in assigning macroscopic thermodynamic properties to small aggregates. In addition, the kinetic treatment of Stranski and Kaischew is only a partial one; the non-equilibrium factor is neglected. Further, there is a possibility that the singular faceted surfaces which they envisaged will become vicinal (see Section D-3) in the presence of an appreciable supersaturation. Also, the derivation of Becker and Doering contains all the limitations of a nearest-neighbor approximation.

One notes that virtually the same statistical mechanical contributions to the free energy of nucleus formation obtain as in the case of droplet nucleation. Finally, there is grave uncertainty about the value of the condensation coefficient α_c and accordingly the same reservations discussed in connection with equation (B-22) for droplets apply to analogous expressions for this case.

(d) *Theory of Homogeneous Nucleation of Droplets from Binary Vapors*

A detailed kinetic analysis of the type leading to equation (B-13) for unary systems has not been performed. However, if the non-equilibrium factor is

* It has been shown by Skapski[105] and Turnbull[62] that the classical expression for the Gibbs free energy of formation of droplet embryos (equation B-1) in the case of critical nuclei is identical with that derived from a nearest-neighbor model if

$$\sigma = -(z_l - z_s)\varepsilon'/2\Omega^{\frac{2}{3}} \qquad \text{(B-26)}$$

where z_l is the molecular coordination number in the interior of the liquid, z_s is the molecular coordination number in a plane liquid–vapor interface, and ε' is the bond energy. This expression should be valid only at 0°K.

ignored, the analog of equation (B-10) for nucleation rate in unary systems is readily obtained. Flood[4,106] showed that, under the usual assumptions, the Gibbs free energy of formation of a critical nucleus from a binary vapor is given by equation (B-4) in which σ is the surface tension and ΔG_v is the volume free energy change of condensation for liquid solution of nuclear composition.* If c is mole fraction of the first component in the critical nucleus and $\bar{\Omega}$ is the average molecular volume of the nuclear solution

$$\Delta G_v = -(kT/\bar{\Omega})[c \ln (p_1/p_{1_e}) + (1 - c) \ln (p_2 p_{2_e})] \qquad \text{(B-27)}$$

where p is the partial pressure of a component in the supersaturated vapor and p_e is the equilibrium vapor pressure of that component over a bulk solution of nuclear composition. The composition of the critical nucleus is readily determined from the condition that the chemical potential of each component must be the same as in the supersaturated vapor. Thus the conditions for equations (B-4) and (B-27) are $(\partial \Delta G^0/\partial r) = 0$, $(\partial \Delta G^0/\partial c) = 0$. The resulting rate expression is identical in form with equation (B-10) for nucleation of droplets from unary vapors. Again, the statistical mechanical contributions to the free energy of nucleus formation are present here, and there is uncertainty as to the value of the condensation coefficient α_c.

3. Theory of Heterogeneous Nucleation

(a) *Theory of Nucleation of Droplets on Gaseous Ions in Unary Vapors*

This very important case was first investigated by J. J. Thomson[4,107] who showed that vapor molecules form stable clusters about all gaseous ions. The Gibbs free energy of formation of a singly charged conductive spherical droplet from the neutral bulk liquid is

$$\Delta G^0 = 4\pi r^2 \sigma + e^2/2r. \qquad \text{(B-28)}$$

Upon differentiating with respect to the number of molecules in the droplet, the chemical potential difference becomes

$$\mu - \mu_0 = kT \ln (p/p_e) = 2\sigma\Omega/r - e^2\Omega/8\pi r^4. \qquad \text{(B-29)}$$

This function exhibits a maximum at a certain value of r, as illustrated in Fig. 7 for the case of water at 265°K, and crosses the abscissa at $r = 4$ Å. This means that even at zero supersaturation the ion is surrounded by a stable shell of vapor molecules. The branch to the left of the maximum in Fig. 7 gives the radius r_1 of these stable droplets as a function of (p/p_e). Of course, a

* A bulk solution having the same composition as the critical nucleus.

finite supersaturation ratio $(p/p_e) > 1$ is required for nucleation of macroscopic droplets. The branch to the right of the maximum in Fig. 7 gives the radius $r^\star$ of the critical nucleus for any imposed supersaturation.

Volmer[4] applied the same analysis and assumptions to the present case as those involved in deriving equation (B-13) for the rate of homogeneous nucleation of droplets. Further, he considered that the spherical aggregates

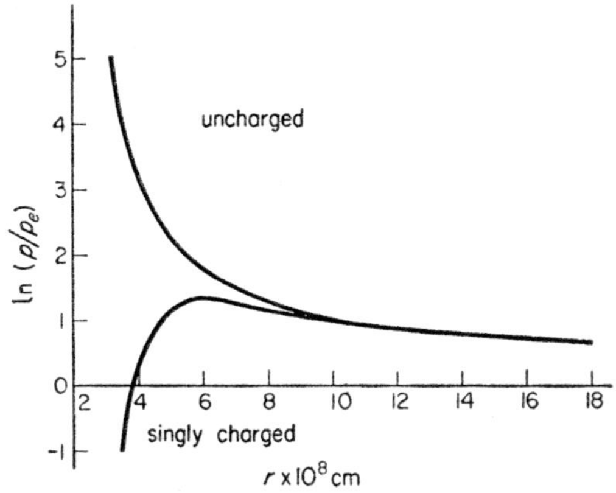

FIG. 7. Metastable-equilibrium embryo size as a function of imposed supersaturation for charged and uncharged clusters. Case of water vapor at 265°K (after Volmer[4]).

are not conductive but have a dielectric constant $\mathscr{E}$. His expression for the Gibbs free energy of formation of a critical nucleus is

$$\Delta G^\star = [\tfrac{1}{3}4\pi\sigma(r^{\star 2} - r_1^2) - 2e^2(1 - 1/\mathscr{E})(1/r_1 - 1/r^\star)] \tag{B-30}$$

and the complete nucleation rate equation is

$$J = \alpha_c Z(4\pi r^{\star 2})p/(2\pi mkT)^{\frac{1}{2}} n_c \exp(-\Delta G^\star/kT) \tag{B-31}$$

where n_c is the concentration of ions and the non-equilibrium factor

$$Z = \{[4\pi\sigma r^{\star 2} - (1 - 1/\mathscr{E})e^2/r^\star]/9\pi i^{\star 2}kT\}^{\frac{1}{2}}. \tag{B-32}$$

Comparing these formulae with equations (B-12) and (B-13) for homogeneous nucleation of droplets in the vapor it is seen that Z is about the same in each case and of the order of 10^{-2}. Also, the concentration of charged clusters n_c is generally much less than the concentration of vapor molecules n, and $\Delta G^\star$ is always less in the present case. Equation (B-31) has all of the defects inherent in (B-13), namely, assumption of macroscopic properties for microscopic aggregates and uncertainty in the condensation coefficient α_c. In addition, another complication arises in connection with the macroscopic assumption: the dielectric constant $\mathscr{E}$ is thought to decrease markedly with droplet size, i.e. with increase in electric field of the central ion, and in a way that is only incompletely understood.[108] However, one notes that the quantum statistical contributions ΔG_s, ΔG_t and ΔG_r (equation B-20) to the free energy of nucleus formation are negligible in this case because of their small dependence on cluster size.

(b) *Theory of Nucleation of Droplets on Foreign Particles* of Subcritical Size in Unary Vapors*

In terms of the usual macroscopic assumptions, Pound and LaMer[109] derived an expression for the Gibbs free energy of formation of a critical cluster about a spherical foreign particle of subnuclear size. If σ_1 is the surface tension of the particle whose radius is r_0 and σ_2 is the interfacial tension between particle and condensing liquid,

$$\Delta G^{\star} = 4\pi r_0^2(\sigma_2 - \sigma_1) - (4/3)\pi r_0^3 \Delta G_v + 16\pi\sigma^3/3\Delta G_v^2. \quad \text{(B-33)}$$

At the present time this is thought to be an unimportant case. Similarly, Reiss[110] and Fletcher[111] have derived expressions for $\Delta G^{\star}$ for nucleation on substrate surfaces of high positive curvature. Again, statistical mechanical contributions are negligible in this case.

(c) *Theory of Nucleation of Droplets by Reactive Impurities*

(1) It is well known[4] that substances such as SO_3, P_2O_5, HNO_3, HCl, etc., when present as impurities in water vapor, bring about nucleation of droplets and fog formation even though the vapor is unsaturated with respect to pure water. It is understood that this phenomenon is due to their effect in greatly lowering the vapor pressure p_e of water, giving rise to a large negative $\Delta G_v = -(kT/\Omega)\ln(p/p_e)$. However, the detailed rate theory of such important processes has not been elucidated.

(2) Certain vapor impurities[4] which are immiscible with the condensing liquid, such as ether or chloroform in water, enhance the nucleation rate because of their effect in reducing the surface tension by adsorption.

4. Experimental Results Relating to Nucleation in the Vapor

(a) *General*

In order to study nucleation of condensed phases in the supersaturated vapor it is necessary to produce a definite known supersaturation. This supersaturation, which for unary systems is often expressed as the ratio of the partial pressure of the vapor to the equilibrium vapor pressure of the

* Immiscible with the condensing liquid.

condensed phase p/p_e, must be produced in such a way that there is no opportunity for its relief by diffusion to the chamber walls. Usually this means that the chilling of the vapor must be done suddenly and this may be accomplished by a sudden expansion of a mixture of the vapor and inert carrier gas. Wilson[112] invented the cloud chamber for this purpose some 65 years ago. It is necessary that the expansion be essentially both reversible and adiabatic to permit computation of the terminal temperature by the formula

$$T_2 = T_1(V_1/V_2)^{[(C_p/C_v)-1]} \tag{B-34}$$

where C_p and C_v refer to molar heat capacities of the gaseous mixture at constant pressure and volume, respectively. A sudden process, of the order of one-tenth of a second in duration, that is also thermodynamically reversible is seldom encountered in a macroscopic system. Nevertheless there is ample evidence[113,114] that the expansion is both reversible and adiabatic at moderately low expansion ratios, $V_2/V_1 = 1.3$ to 1.4, corresponding to temperature reductions of 20 to 30°C. Makower[113] demonstrated experimentally that equation (B-34) is valid for moderate expansions of air. Flood[4,114] used an oscilloscope to record the temperature, as sensed by a resistance thermometer, during the expansion of dry air in a container of about 1 liter capacity. Varying the speed of the expansion to give durations between 1/100 and 2/10 sec had no effect on the terminal temperature and he concluded that expansions of these speeds are essentially reversible. He also studied the effect of the degree of the expansion on the adiabaticity. Small expansion ratios of the order of $V_2/V_1 = 1.2$ gave terminal-temperature durations of the order of 1 sec. However, expansion ratios greater than 1.4 resulted in negligible duration of the terminal temperature. Thus expansion ratios greater than 1.4 are not adiabatic in chambers of the usual size ($\sim$1 liter). In any case if it is established that the expansion is both reversible and adiabatic, the supersaturation at the terminal temperature can be easily computed from the partial pressure of vapor at the initial temperature, the expansion ratio and the vapor pressure of the condensate at the terminal temperature.

The best test of nucleation theory would be to measure the actual number N of droplets formed during the period Δt of a given terminal temperature and determine J as $N/\Delta t$. However, as indicated above, the period Δt is short and difficult to determine precisely by either measurement or calculation. Also, for a period of time which is of the same order as Δt, the expanded system is in a temperature range not far above T_2. More importantly, the period of growth of the nuclei is not long compared with Δt, and so the vapor is being continuously depleted in a very complicated fashion as nucleation proceeds. Finally, the heat of condensation is sufficient to elevate the temperature of the gas appreciably; a water droplet of 10^{-2} cm diameter liberates enough heat to raise the temperature of 1 cc of air by 1°C. Accordingly the supersaturation is neither constant nor known during the brief period at the terminus of the expansion. For these reasons, in the opinion of the present authors, almost all measurements of actual nucleation frequency in precipitation from the vapor are without precise quantitative significance. Fortunately, the dependence of nucleation rate on supersaturation in the various kinetic expressions is so great that one may specify a critical supersaturation below which nucleation rate is negligible and above which it is very high. This theoretical prediction is universally borne out in experimental investigations; the change in supersaturation required to produce or eliminate observable precipitation is usually beyond the precision of the experiment. For these reasons most meaningful data on nucleation in the vapor are reported as critical supersaturation ratios, $(p/p_e)_{\mathrm{cr}}$, where $\ln J = 0$.

Volmer and Flood[4,85] have shown that not all vapors may be studied by expansion methods. It is readily shown that the supersaturation ratio obtained by expansion of a saturated vapor at T_1 to terminal temperature T_2 in a piston cloud chamber is given by

$$\ln (p/p_e) = (\Delta H_{\mathrm{vap}}/R) \cdot \frac{T_1 - T_2}{T_1 T_2} - \frac{C_p/C_v}{(C_p/C_v) - 1} \ln (T_1/T_2). \qquad \text{(B-35)}$$

On the other hand, as will be shown in the following discussion, from equations (B-2) and (B-13)

$$\ln (p/p_e)_{\mathrm{cr}} = L(\sigma/T_2)^{\frac{3}{2}} \qquad \text{(B-36)}$$

where L is approximately constant. In many cases the curves of these equations fail to intersect and thus even arbitrarily high expansions will not produce condensation. One notes that the high surface tension of metals ($\sigma \simeq 10^3$ ergs/cm^2) precludes study of their vapors by the cloud chamber techniques.

Another general experimental consideration has to do with the role of gaseous ions in the nucleation of droplets. As described in Part 3 of the present section, vapor molecules attach themselves to simple ions to form stable charged clusters of considerable size. Hazen[115] and Pollermann and Stranski[116] have measured the mobility of these clusters and found, as would be expected, that it is much less than for the simple ions in a dry carrier gas. Further, these charged clusters are continuously being produced in the carrier gas by cosmic rays and natural radioactivity. The rate in air is several ion pairs per cubic centimeter per second, leading to a steady-state concentration of about 1000 ion pairs per cc. Accordingly extreme caution must be exercised in assuming that a given applied electric field renders a cloud-chamber volume ion-free.

The cloud chamber may be of the piston or diaphragm type or it may be simply a glass vessel which can be opened to a vacuum reservoir. Most of the critical supersaturation data have been obtained with cloud chambers of these designs.[117,118] However the vapor-jet technique[119] has been employed in a number of investigations.[120,121] Also, the diffusion cloud chamber[122] has proven useful in phase transition studies, particularly in connection with determination of critical supercoolings for solidification of liquids.[123]

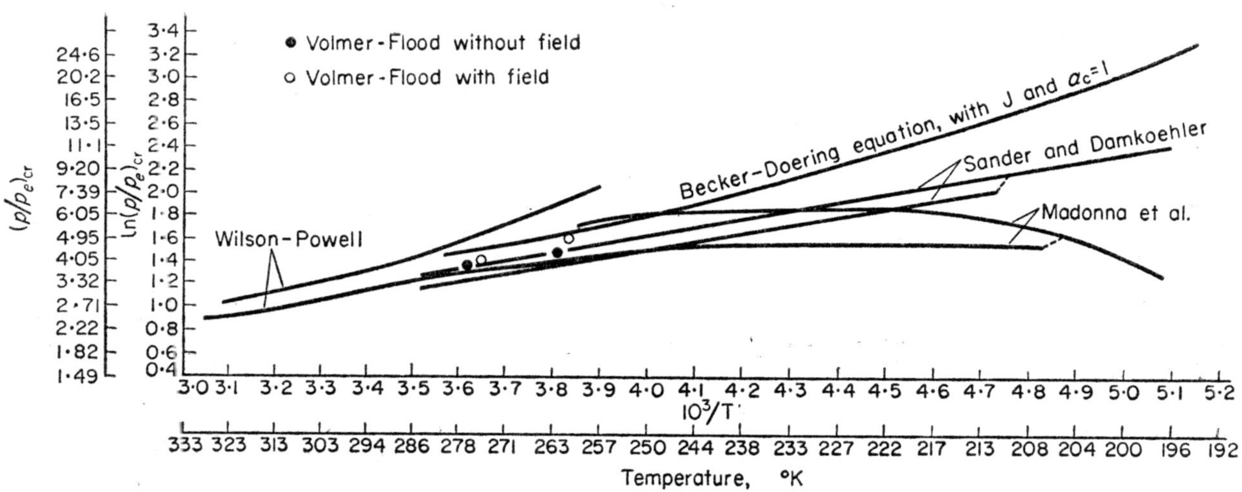

FIG. 8. Assembled critical supersaturation data for water vapor.

(b) *Experimental Results on Homogeneous Nucleation of Droplets from Unary Vapors*

In the opinion of the authors, some of the most significant quantitative measurements of critical supersaturation for homogeneous nucleation of droplets are those of Wilson[88–90,112] and Powell[91] who investigated the system water vapor in air by a piston-chamber technique. In this work, dust particles were removed from the air by several repeated expansions. In some of his later experiments Wilson studied the effect of gaseous ions by using X-rays to produce them and electric fields to diminish their concentration. For a terminal temperature of 268°K he found that the critical supersaturation ratio $(p/p_e)_{cr}$ for nucleation of water droplets on negative ions, the first "rain limit", is 4.1. The corresponding critical supersaturation ratio for nucleation on positive ions, the second "rain limit", was determined to be about 6. For a terminal temperature of 255°K Wilson found that the critical supersaturation ratio for nucleation of a dense fog of water droplets is 8.0. He identified this "fog limit" as the threshold for onset of homogeneous nucleation of droplets.

Powell,[91] using the fog-limit criterion for homogeneous nucleation, extended the measurements on water vapor to cover a range of temperature. The assembled data for water vapor are presented in Fig. 8. It is seen that neither the absolute value of $(p/p_e)_{cr}$ nor its temperature dependence are in good agreement with the predictions of the Becker–Doering equation (B-13) with J and $\alpha_c = 1$. The most that may be said is that equation (B-13) affords a semi-quantitative description of the data. In view of the inadequacies of equation (B-13) which were described in the section on Homogeneous Nucleation Theory, such agreement as there is must be regarded with reservation.

Inasmuch as the pre-exponential terms are relatively independent of temperature and supersaturation, equation (B-13) may be written for the critical condition as

$$J_{cr} = \alpha_c B \exp(-\Delta G^{\star}/kT), \tag{B-37}$$

and from equation (B-2)

$$\ln (p/p_e)_{cr} = L(\sigma/T)^{\frac{3}{2}} \tag{B-38}$$

in which

$$L = [16\pi\Omega^2/3k^3 \ln(\alpha_c B/J_{cr})]^{\frac{1}{2}}. \tag{B-39}$$

From the above discussion it seems unlikely that J_{cr} should be taken as appreciably different from unity, at least for the experimental conditions under present consideration. Thus Farley[62,124] concluded that if α_c is independent of temperature and supersaturation, L should be approximately constant and hence the factor $(T/\sigma)^{\frac{3}{2}} \ln (p/p_e)_{cr}$ should not vary appreciably. In an analysis of Powell's data, Farley[124] showed that for water this factor is approximately constant at 11 ± 10% over a temperature range of some 70°C.

Since equation (B-13) has been shown to be incomplete, interpretation of this finding is difficult. However, one notes that the statistical mechanical contributions $\overline{\Delta G_g} = \Delta G_s + \Delta G_t + \Delta G_r$ (equation B-20) to $\Delta G^{\star\prime}$ are relatively independent of temperature. Thus Farley's result should hold with the modified equation (B-22) if the constant is redefined as

$$L = 16\pi\Omega^2/3k^3 \ln(\alpha_c B) \cdot [1 - (\overline{\Delta G_g}/kT \ln \alpha_c B)]^{\frac{1}{2}} \tag{B-40}$$

In the past, the apparent discrepancy between theory and experiment has been expressed in terms of the value of surface tension required to make equation (B-13) fit the data.[62,125] For the case of water an increase of 13% over the measured surface tension σ was needed. However, as discussed in the section on Homogeneous Nucleation Theory there are other aspects in which the Becker–Doering equation (B-13) is incomplete and thus this procedure may also be inadequate.

There are not many critical supersaturation data on vapors other than water which may be represented as pertaining to homogeneous nucleation. Clarke and Rodebush[126] used helium instead of air to study the nucleation of droplets from water vapor and from benzene vapor. They took the onset of "appreciable" droplet precipitation rather than fog formation as the criterion of critical supersaturation and found that for water the observed critical supersaturation ratio is 7 at 266°K, which is only slightly less than Wilson's fog limit. Evidently the ion concentration in their pure helium-water vapor mixture was sufficiently low to elevate the ion limits almost to the fog limit in accordance with the prediction of equation (B-31). However, this result is difficult to understand in view of the facts that there was about 1 mole percent of water vapor in the mixture and the first ionization potential of a water molecule is nearly the same as for an oxygen molecule, i.e. 12.6 volts.* Their observed critical supersaturation ratio for benzene is 8 at 250°K and corresponds to a rate about 10^6 times greater than predicted by equation (B-13) with J and $\alpha_c = 1$. Again, in the opinion of the present authors, this observation represents heterogeneous nucleation on some attenuated concentration of gaseous ions.

O'Konski and Higuchi[120] used a turbulent-jet technique to measure the steady-state production rate of droplets given by

$$I = \int J(p/p_e, T)\, \mathrm{d}v' \tag{B-41}$$

where the integration is carried out over that volume v' of the jet in which nucleation is occurring. They were able to determine separately the coefficient and exponential factors of equation (B-13). The results[121] for dibutyl phthalate were in remarkable accord with the Becker–Doering theory. Only fair agreement was obtained with n-octadecane and sulfur, while the agreement

* In this connection one notes that a reduction of ion concentration n_c by a hundred-fold would not appreciably elevate $(p/p_e)_{\mathrm{cr}}$.

in the case of triethylene glycol was poor. The present authors note that the technique of Higuchi and O'Konski is rather involved. Further, it seems that no effort was made to assess the role of gaseous ions in the nucleation process. In view of the concentration of droplets observed ($1 - 10^3$ cc^{-1}), heterogeneous nucleation on gaseous ions would appear to be a distinct possibility.

Recently, Edwards and Woodbury[127] measured the critical supersaturation ratio for nucleation of droplets from pure helium vapor by means of an expansion technique. Their results are much lower than those predicted from equation (B-13) with α_c and $J = 1$. The present authors believe that their data represent heterogeneous nucleation on ions or perhaps even dust particles.

(c) *Experimental Results on Homogeneous Nucleation of Crystals from Unary Vapors*

Sander and Damkoehler[128] observed that the precipitate at the rain limit of water vapor in air changed in appearance from that of spherical droplets to that of acicular ice crystals at a terminal temperature of -61°C and a critical supersaturation ratio of about 8.* Further, the effect of an electric field disappeared below this temperature. Madonna *et al.*[129] found the same effects for water vapor in nitrogen, except that the apparent transition from droplet to crystal nucleation occurred at -65°C and a critical supersaturation ratio of about 5. The data are summarized in Fig. 8. Tentatively, these results are interpreted to mean that ions or charged clusters of water molecules of sub-nuclear size do not promote heterogeneous nucleation of ice crystals and, below -65°C, the homogeneous nucleation of ice crystals from the vapor is faster than the heterogeneous nucleation of water droplets on ions. Evidently the charged clusters of water molecules do not have a lattice structure which resembles that of ice. However, it seems certain that all of the spherical precipitate particles observed at terminal temperatures lower than -41°C are ice balls because it has been firmly established that this is the critical temperature for rapid homogeneous nucleation of ice from supercooled water.[130]

Maybank and Mason[131] prefer to interpret these results in terms of a change in the growth mechanism of frozen droplets at -65°C rather than as a change in the nucleation mechanism.

It is interesting to note that pure tertiary butyl alcohol (f.p. 25°C) exhibits no transition in mode of nucleation down to a terminal temperature of -75°C.[132]

Finally, one notes that the experimental data on homogeneous nucleation of crystals from the vapor are too sparse to permit meaningful quantitative comparison with theory.

* With respect to water.

(d) *Experimental Results on Homogeneous Nucleation of Droplets from Binary Vapors*

To the best of the authors' knowledge, the only critical expansion data for droplet formation from binary vapors which may be said to represent homogeneous nucleation are those of Needels[133] who used the Wilson technique and fog-limit criterion to investigate two compositions of ethanol and water in air over a range of temperature. His raw data appear to be sound but much work remains to be done in comparing them with the predictions of the theory or with other data.

(e) *Experimental Results on Heterogeneous Nucleation of Droplets on Ions in Unary Vapors*

Volmer[4] and his school believed that the Wilson fog limit was not an adequate criterion for the onset of homogeneous nucleation of droplets from vapor on the seemingly reasonable grounds that precipitation on ions would elevate the apparent critical supersaturations for homogeneous nucleation by depleting the vapor concentration and raising the actual temperature above the calculated terminal temperature. They proposed as a proper criterion the incidence of appreciable droplet formation in the presence of an electric field which was presumed to remove all ions and charged clusters. In other words they measured the rain limit in the presence of an electric field. As shown in Table 2, these measured critical supersaturations for water and a number of organic vapors are in remarkable agreement with the predictions of the Becker–Doering equation (B-13) for homogeneous nucleation of droplets

TABLE 2

COMPARISON OF MEASURED AND CALCULATED VALUES OF CRITICAL SUPERSATURATION RATIO. DATA OF VOLMER AND FLOOD[85]

Vapor	Number of molecules in stable nucleus	Nucleus radius r in angstroms	$(p/p_e)_{cr}$ calculated	$(p/p_e)_{cr}$ measured
Water, 275.2°K	80.0	8.9	4.2	4.2 ± 0.1
Water, 261.0°K	72.0	8.0	5.0	5.0
Methanol, 270.0°K	32.0	7.9	1.8	3.0
Ethanol, 273.0°K	128.0	14.2	2.3	2.3
n-propanol, 270.0°K	115.0	15.0	3.2	3.0
Isopropyl alcohol, 265.0°K	119.0	15.2	2.9	2.8
n-butyl alcohol, 270.0°K	72.0	13.6	4.5	4.6
Nitromethane, 252.0°K	66.0	11.0	6.2	6.0
Ethyl acetate, 242.0°K	40.0	11.4	10.4	8.6 to 12.3

with α_c and $J = 1$. Accordingly, for some years this agreement was accepted as a firm basis for homogeneous nucleation theory.

Since it appears that equation (B-13) is quite incomplete, as discussed above, one must regard such quantitative agreement as purely fortuitous. Further, there is good reason to think that the Volmer and Flood(85) data represent heterogeneous nucleation on ions.

Sander and Damkoehler(128) and Madonna *et al.*(129) used the Volmer criterion to measure the critical supersaturation ratio of water vapor over a wide range of temperature. The results are presented in Fig. 8 where the upper curve of each set of observations represents data taken in the presence of an electric field and the lower curve describes data taken in the absence of a field. It is seen that the two curves of each set of observations are parallel, indicating that the temperature dependence of the critical supersaturation is about the same with or without an applied field. According to any reasonable theory the upper curve should have an appreciably steeper slope, consistent with a higher activation energy for homogeneous nucleation.* The fact that it does not indicates that it represents nucleation on ions or charged clusters, just as does the lower curve. That the two curves do not coincide is attributed to the lower concentration of charged clusters in the presence of an electric field. Also one notes that the ordinate distance between the two curves is greater for the data of Madonna *et al.* than for those of Sander and Damkoehler. This is thought to be due to the fact that ion removal was more effective in the former work.† In other words, experience seems to show that the more effective the removal of charged clusters the more closely the Volmer rain limit approaches the Wilson fog limit.

The Volmer rain limit is characterized by a droplet concentration N of the order of 1 per cc. Thus in view of the fact that the steady-state ion concentration $\bar{n}_{c_0}$ in air in the absence of a field is approximately 10^3 pairs per cc, it would appear that the reduction in ion concentration occasioned by an applied electric field is only about a thousand fold. Also, one may show from equation (B-31) that the ion concentration would have to be reduced by a factor of the order of 10^6 before the critical supersaturation for nucleation on ions would approach that for homogeneous nucleation. Therefore one might, at first sight, question that ions are responsible for nucleation at the Volmer rain limit. However, it is noted that the rate of production $\dot{n}_c$ of ions in air from natural high-energy radiation is roughly 10 cc^{-1} sec^{-1}. If the terminal temperature endures for a time Δt (of the order of 1/10 second) and $\bar{n}_c$ is the steady-state concentration of ions in the field, the expected maximum concentration of droplets is $n_m = \bar{n}_c + \dot{n}_c \Delta t$. The period during which droplets may be formed on $\bar{n}_c$ is Δt while the corresponding period for $\dot{n}_c \Delta t$ is, very

* Note that this difference is indeed found in the Wilson–Powell results of Fig. 8.

† The difference between the two sets of results at low temperatures is thought not to be due to the fact that Madonna *et al.* used nitrogen instead of air as the carrier gas. Possible reasons for this discrepancy are discussed in reference 129. One notes the agreement on the negative ion limit at high temperatures.

approximately, τ, the reciprocal of the velocity $\dot{x}$ of the charged clusters in the applied field. Thus expressing equation (B-31) as $J = n_c(\partial J/\partial n_c)$, the observed concentration of droplets will be

$$N \cong (\partial J/\partial n_c)\Delta t(\bar{n}_c + \dot{n}_c\tau) \qquad \text{(B-42)}$$

for $\tau < \Delta t$. On the other hand, in the absence of a field the steady-state concentration n_{c_0} is much greater than $\dot{n}_c\Delta t$ and

$$N \cong n_{c_0}(\partial J/\partial n_c)\Delta t. \qquad \text{(B-43)}$$

Hence for a fixed Δt, the increase in $(\partial J/\partial n_c)$ required to give a certain value of N upon application of a field is $n_{c_0}/(\bar{n}_c + \dot{n}_c\tau) \cong n_{c_0}/\dot{n}_c\tau$ under common experimental conditions. For the usual values of applied field $\dot{x}$ may be of the order of 10^3 cm/sec. Therefore it is seen that increases in $(\partial J/\partial n_c)$ of the order of 10^6 may be necessitated by application of a field, and thus the observations are interpreted as condensation on ions. Further, it is evident from this demonstration that ions are never completely removed by applied electric fields, and only very high values of the field will cause the Volmer rain limit to approach the Wilson fog limit.

For these reasons the present authors are inclined to reject Volmer's criticism of the Wilson fog limit as a criterion of homogeneous nucleation. Rather they tend to reject the Volmer rain limit as a criterion for anything but nucleation on charged clusters. Further, pending a similar detailed investigation of the organic vapors listed in Table 2, they choose to classify the Volmer and Flood results for these vapors as descriptive of heterogeneous nucleation on ions and for this reason have included them in the present section.

One notes that equation (B-31) for heterogeneous nucleation of droplets on ions could be used to describe the data of Table 2 by selecting appropriate values for the three adjustable parameters: the concentration of ions n_c, the dielectric constant $\mathscr{E}$ and the condensation coefficient α_e. Similarly, those data of Fig. 8 which are interpreted as representing nucleation on ions (all except the Wilson–Powell fog-limit data) could be described in terms of equation (B-31). However in view of the multiplicity of adjustable parameters in the present theoretical approach, the authors feel that such curve-fitting would not be particularly fruitful. As an illustration, it is found that the negative-ion limit for water vapor of 4.1 at 0°C is described by equation (B-31) with α_c and $J = 1$, $n_c = 10^3$ cc^{-1} and a dielectric constant $\mathscr{E} = 1.85$ (versus $\mathscr{E} = 80$ for bulk water).

Finally, a comment is in order on the difference in catalytic potency for droplet nucleation between ions of opposite sign. We have seen from Wilson's work that the critical supersaturation ratio for nucleation of water droplets on the normal concentration of negative ions in air is 4.1 at 0°C while for positive ions it is about 6. Laby[134] and Scharrer[135] investigated various vapors in air and found that most organic vapors exhibit preferential condensation on positive ions while benzene and CCl_4 show no preference. Of the vapors investigated, only water and chloroform condense more rapidly on negative ions. These differences are qualitatively explained by assuming that adsorption at the surface of the clusters produces an electrical double layer.(136) If the sign of the field due to this electrification is the same as that of the ion, there is a reduction in free energy of formation of the critical nucleus.

(f) *Experimental Results on Heterogeneous Nucleation of Droplets on Ions in Binary Vapors*

Using the Volmer rain-limit criterion for homogeneous nucleation, which as we saw in part (e) above is the rain limit in the presence of an electric field, Flood(106) measured the critical supersaturation ratio for the system ethanol–water in air over the entire range of composition. For the reasons outlined above, the present authors tentatively prefer to classify these data as representative of heterogeneous nucleation on an attenuated concentration of ions. The agreement with the earlier homogeneous nucleation theory for binary

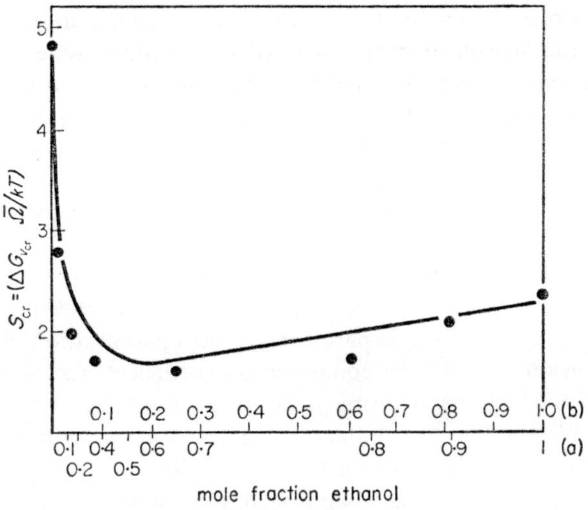

(a) in cloud-chamber liquid
(b) in critical nuclei

FIG. 9. Critical supersaturation ratio for nucleation of droplets in the presence of an electric field for the system ethanol–water in air. Measurements by the Volmer rain-limit criterion. Solid curve is from homogeneous nucleation theory. Points are data, results of Flood.(106)

vapors is good, as shown in Fig. 9 and Table 3. Again this result appears questionable because the same factors which modify equation (B-13) to give equation (B-22) apply to this case.

TABLE 3

CRITICAL SUPERSATURATION DATA FOR NUCLEATION OF DROPLETS IN THE PRESENCE OF AN ELECTRIC FIELD FOR THE SYSTEM ETHANOL–WATER IN AIR

Measurements by the Volmer rain-limit criterion. Calculated values from earlier homogeneous nucleation theory. Results of Flood.(106)

Liquid composition mole % ethanol	Terminal T °K	Nuclear composition mole % ethanol	S_{cr}, experimental	S_{cr}, theoretical
0.0	263.7	0.0	4.85	4.85
3.86	272.8	0.8	2.64	3.3
11.5	275.6	3.7	1.97	2.45
35.3	280.4	8.5	1.75	1.92
67.1	277.0	25.0	1.62	1.70
77.9	274.9	61.0	1.77	1.97
90.3	273.8	82.0	2.07	2.16
100.0	273.2	100.0	2.34	2.30

(g) *Experimental Results on Nucleation of Droplets or Crystals on Foreign Particles of Subcritical Size*

The authors know of no quantitative measurements of this kind. However there is an extensive literature relating to nucleation on foreign particles of *supercritical* size by cloud-chamber techniques.* This will be discussed in the next section on Heterogeneous Nucleation on Substrates, because the theory developed there is more appropriate for this case.

(h) *Experimental Results on Nucleation of Droplets by Reactive Impurities*

The data found in the rather voluminous literature on this subject are sufficiently qualitative to discourage analysis of the results. Accordingly this topic will not be treated here. In passing however, one should call attention to the interesting phenomena found in cloud-chamber studies of vapor–gas mixtures which have been irradiated with ultraviolet light. The ultraviolet light evidently produces a variety of condensation nuclei, most of which are as yet unidentified.(137)

5. CONCLUSION

Important statistical mechanical contributions(65) to the free energy of formation of embryos in homogeneous nucleation have in the past been

* In this connection attention is called to the very interesting work of LaMer and co-workers(526,527) who used the principles discussed here to produce monodisperse aerosols of supercooled liquid sulfur droplets.

neglected. The sum of these contributions (equation B-20) is large and negative, and the net effect is to increase the nucleation rate predicted by the Becker–Doering equation (B-13) by a factor of about 10^{17}. This destroys the long-standing agreement between theory and critical supersaturation data from cloud-chamber work for the critical supersaturations calculated from the modified equation (B-22) are now too low by a factor of about 1.3. However the fact that the observed critical supersaturations may be correlated with the physical properties of the liquids lends some credence to the form of equations (B-4), (B-13) and (B-22). It is suggested that the problem should be treated as a transient, considering thermal non-accommodation in the formation of the embryos. Such a treatment might yield an effective condensation coefficient α_c sufficiently less than unity to restore agreement with observation.

Efforts to evolve a more precise theoretical expression for homogeneous nucleation by removing the gross assumption inherent in assigning macroscopic thermodynamic properties to clusters involving only a few molecules have thus far met with limited success.

Evidently the Wilson fog limit is the only trustworthy experimental criterion for critical supersaturation in homogeneous nucleation of droplets from the vapor. Hence the critical supersaturation ratios obtained by Wilson and Powell(91) for water vapor in air seem to be virtually the only quantitative data representative of homogeneous nucleation.

It now appears that most cloud-chamber data, including those of Volmer and Flood,(4,85) represent heterogeneous nucleation on ions or charged clusters. Unfortunately, equation (B-31) the theoretical rate expression for heterogeneous nucleation on ions contains too many adjustable parameters for meaningful comparison with the data.

C. HETEROGENEOUS NUCLEATION ON SUBSTRATES

1. Introduction

The basic ideas developed in the last section, which dealt with nucleation in the vapor phase, will now be extended and applied to heterogeneous nucleation on substrates. In succession, the theory will be developed for nucleation on (i) flat clean substrates, (ii) clean surfaces which are not flat and (iii) dirty surfaces. Finally, experimental results will be summarized and compared with theory. Review articles dealing with the subject of this section have been written recently by Pound,[63] Devienne,[12] Datz and Taylor,[138] Ehrlich[15] and Hirth.[87]

2. Theory

(a) *Heterogeneous Nucleation of Cap-Shaped Nuclei of Spherical Symmetry on Clean Flat Substrates*

1. *Constant Population of Adatoms*

(a) *Classical treatment.* As shown in Part A, molecules or atoms impinging on a substrate to which they have strong binding forces will stick and thermally equilibrate with that substrate. The relaxation time for the flux of atoms evaporating from a metastable adsorbed population n_s to become equal to the impingent flux is about the same as the residence time τ_s (equation A-8). In many cases τ_s is very short compared to the observation time in a nucleation experiment so that n_s may be considered constant.* In this circumstance[139] the impingent flux (equation A-1) will equal the vaporization flux:

$$p/(2\pi mkT)^{\frac{1}{2}} = (n_s/\tau_s) = n_s\bar{\omega}\exp(-\Delta G^{\star}_{\text{des}}/kT) \qquad \text{(C-1)}$$

where p is the equivalent vapor pressure at surface temperature T that would give the true impingent flux as determined by experimental conditions. Here the evaporation flux is the product of n_s and $(1/\tau_s)$, the frequency of evaporation of an adatom.† $\Delta G^{\star}_{\text{des}}$ is a free energy of activation for desorption of an

* Note that this steady-state concentration of adatoms exists only until stable nuclei begin to form. Once stable growing nuclei form they will rapidly deplete the population of adsorbed atoms.

† Equations (A-8) and (C-1) follow from the Wert–Zener[140] treatment of absolute rate theory. Thus $\bar{\omega} = (kT/h)\,[1 - \exp(-h\nu/kT)]$ and at temperatures greater than the Debye temperature $\bar{\omega} \simeq \nu$, the Einstein frequency. Some justification for the use of this expression for $\bar{\omega}$ is given in part (D-1).

adatom. It follows from equation (C-1) that

$$(p/p_e) = (n_s/n_{s_e}) \tag{C-2}$$

as long as the monomer is the predominant species in both the vapor and adsorbed states.

As considered in the discussion preceding equation (B-8), embryos of various sizes on the surface are assumed to be in metastable equilibrium with adatoms. In particular, the concentration of nuclei of critical size is roughly approximated* by

$$n^\star = n_s \exp(-\Delta G^\star/kT) \tag{C-3}$$

where $\Delta G^\star$ is the free energy of formation ignoring certain statistical contributions. If the nucleus has an isotropic specific surface free energy σ_{c-v}, it

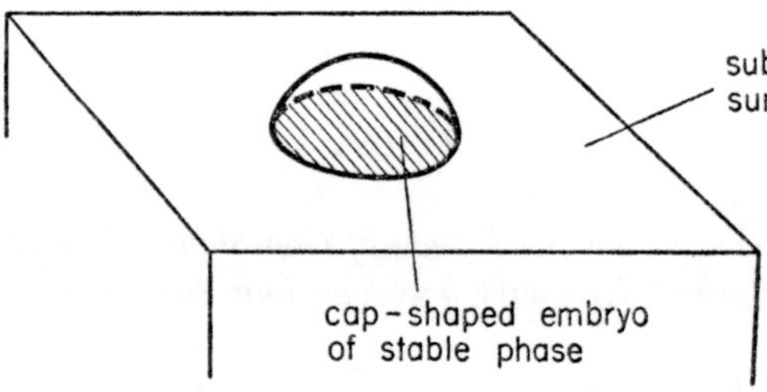

FIG. 10. Schematic illustration of a cap-shaped embryo.

will have the shape of a segment of a sphere as shown in Fig. 10. The equilibrium contact angle θ is related to σ_{x-v} and σ_{c-x}, respectively the substrate–vapor and condensate–substrate specific interfacial free energies, by the Young[141] equation

$$\sigma_{x-v} = \sigma_{c-x} + \sigma_{c-v} \cos\theta.\dagger \tag{C-4}$$

Referring to Fig. 10, the free energy of formation of an embryo containing i atoms is

$$\Delta G = \pi r^2 \sin^2\theta(\sigma_{c-x} - \sigma_{x-v}) + 4\pi r^2 \phi_1(\theta)\sigma_{c-v} + (4\pi/3) r^3 \phi_2(\theta)\Delta G_v \tag{C-5}$$

in which r is the radius of the sphere, ΔG_v is given by equation (B-2) and ϕ_1 and ϕ_2 are the appropriate geometric functions of θ. Maximizing equation (C-5) with respect to r, one obtains[4] the free energy of formation of the nucleus of critical size:

$$\Delta G^\star = (16\pi\sigma_{c-v}{}^3/3\Delta G_v^2)\phi_3(\theta),\ \phi_3(\theta) = (2 + \cos\theta)(1 - \cos\theta)^2/4. \tag{C-6}$$

* A statistical mechanical correction[65] factor analogous to equation (B-23) will be given in the following.

† One notes that this relationship has been rigorously derived by Gibbs.[68]

Hence if the equilibrium contact angle θ is zero, i.e. if there is complete "wetting" of the substrate by the condensate, $\Delta G^\star$ equals zero and nucleation will be most rapid. If $\theta = 180°$ corresponding to no "wetting" at all, $\phi_3(\theta) = 1$ and equation (C-6) reduces to equation (B-4), the expression for homogeneous nucleation. This is the limiting case in which the substrate has negligible potency as a nucleation catalyst. $\phi_3(\theta)$ as a function of θ is plotted in Fig. 11.

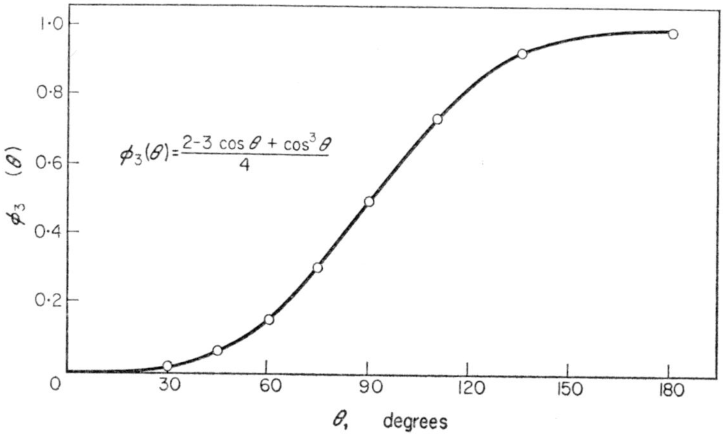

FIG. 11. ϕ_3 (θ) plotted as function of the contact angle θ.

Equations (C-5) and (C-6) neglect the contribution $2\pi r^\star \varepsilon_1 \sin\theta$ of the line tension at the periphery of the cap to the free energy of formation of the embryo or nucleus. In the case of small nuclei ($r^\star \simeq 10^{-7}$ cm) it could be of the same order of magnitude as the surface term. Further, if the line tension is important, the equilibrium shape will not be a segment of a sphere and equation (C-4) will be invalid. However, this complication is ignored in the present treatment, because it will emerge that the available data are insufficiently precise to determine if this contribution is important. In other words, the existing data may be just as well described by either formulation and hence the simpler is chosen. However, future improvements in experimental technique may one day justify use of the more detailed analysis.

The rate of nucleation

$$J = Z\omega n^\star \tag{C-7}$$

where ω is the frequency with which a single atom joins a critical nucleus to promote it to a stable growing nucleus and Z is a non-equilibrium factor (cf. equation B-12). This process may occur by either direct addition from the vapor[60] or surface diffusion of an adatom.[142] However, Pound *et al.*[142] have shown that the latter process is more rapid by a factor exp $[(\Delta G^\star_{\text{des}} - \Delta G^\star_{\text{sd}})/kT]$ where $\Delta G^\star_{\text{sd}}$ is the free energy of activation for surface diffusion. In general the surface-diffusion process will be predominant. Thus the frequency ω is given by the product of $n_s 2\pi r^\star a \sin\theta$, the probability that an adatom is adjacent to a critical nucleus, and $\bar{\omega}\exp(-\Delta G^\star_{sd}/kT)$, the frequency with which an adjacent adatom will jump to join the nucleus.* Here $r^\star$ is the radius of the nucleus, and a is the diffusion jump distance. In this case of heterogeneous nucleation on a substrate the non-equilibrium factor is

$$Z = (\Delta G^\star/3\pi kTi^{\star 2})^{\frac{1}{2}}. \tag{C-10}$$

Substituting in equation (C-7) and reducing the resulting expression(142) one obtains:

$$J = C_1' p^2 \exp\left[(2\Delta G^\star_{des} - \Delta G^\star_{sd} - \Delta G^\star)/kT)\right] \tag{C-11}$$

where† $C_1' = [a \sin\theta \ln(p/p_e)/8\pi\bar{\omega}m][3/\phi_3(\theta)\sigma_{c-v}kT]^{\frac{1}{2}}$.

Now the nucleation rate is so sensitive to the ΔG_v term in $\Delta G^\star$ of equation (C-11) (see equation C-6) that there is in essence a critical value $\Delta G_{v_{crit}} = -(kT/\Omega)\ln(p/p_e)_{crit}$ where $\ln J = 0$.‡ Above this critical supersaturation $J \simeq \infty$ and below it $J \simeq 0$. Accordingly,(142) setting $\ln J = 0$ and rearranging equation (C-11), it emerges that

$$(1/\Delta G_{v_{crit}})^2 = \left[\frac{3}{16\pi\sigma_{c-v}{}^3\phi_3(\theta)}\right][kT(\ln C_1' + 2\ln p) + 2\Delta G^\star_{des} - \Delta G^\star_{sd}] \tag{C-12}$$

Thus if one measures $AG_{v_{crit}}$ and plots $(1/\Delta G_{v_{crit}})^2$ as a function of $kT(\ln C_1' + 2\ln p)$ a straight line should result. In this case $\sigma_{c-v}{}^3\phi_3(\theta)$ may be obtained from the slope and $2\Delta G^\star_{des} - \Delta G^\star_{sd}$ from the intercept.

(b) *Recent modifications.* As will be shown in the part dealing with Experiment, this approximate relationship (equation C-12) affords a rough description of the data. However, it neglects statistical mechanical corrections to the free energy of formation of the critical nucleus which were recently introduced by Lothe and Pound.(65) In the development leading to the free energy requirement ΔG_t (equation B-18) for energizing the critical droplets in homogeneous nucleation from the vapor it was necessary to consider the

* As long as the product of this surface-diffusion jump frequency and the jump distance a is less than $(2kT/m)^{\frac{1}{2}}$ the adsorbed atoms may be regarded as being localized(143) such that diffusion occurs by discrete jumps. This condition reduces essentially to the requirement that

$$\Delta G^\star_{sd} > kT, \tag{C-8}$$

which should hold for solids. For the case of deposition of liquids this condition may not obtain. Then it would be more appropriate to regard the adsorbed atoms (or molecules) as translating in a two-dimensional gas and the frequency ω would be the product of $2\pi r^\star \sin\theta$, the circumference of the critical nucleus, and $(n_s/\pi)(\pi kT/2m)^{\frac{1}{2}}$, the collision frequency in a two-dimensional gas, or

$$\omega = 2r^\star n_s \sin\theta(\pi kT/2m)^{\frac{1}{2}} \tag{C-9}$$

† For a given system C' is approximately constant over a range of p and T and typically assumes a value of about 10^7 to 10^8 cm^2 dyne^{-2} sec^{-1}.

‡ The critical value of J is in essence the sensible value that can be measured experimentally. For experiments where nuclei are observed optically, this corresponds to $J \simeq 1$ nucleus/cm^2 sec or $\ln J \simeq 0$. For other cases the critical value will differ; e.g. when nucleation is observed by field emission microscopy the critical value of J is about 10^8 nuclei/cm^2 sec.

complexions involved in distributing the nuclei among cells in the translational phase space. There is an analogous problem in nucleation upon substrates. For the case in which adsorbed molecules and clusters are localized[143] as assumed in the discussion preceding equation (C-10), one must consider the number of complexions that arise from distributing adsorbed molecules and crystalline embryos among the n_0 sites on a solid substrate lattice.* Taking the standard state of the embryos as pure and equal in concentration to adsorbed monomer n_s, noting that the concentration of adsorbed monomer greatly exceeds the concentration of sites occupied by all sizes of clusters, and invoking Fermi–Dirac statistics, the number of complexions is

$$W = n_0!/n_s!(n_0 - n_s)! \qquad \text{(C-13)}$$

and the contribution to the free energy of formation is

$$\Delta G_q = (kT/n_s)\{n_0 \ln [(n_0 - n_s)/n_0] + n_s \ln [n_s/(n_0 - n_s)]\} \qquad \text{(C-14)}$$

In the usual experimental case $n_0 \gg n_s$ and

$$\Delta G_q = -kT \ln (n_0/n_s) \qquad \text{(C-15)}$$

from which the factor of increase in the equilibrium concentration of nuclei is n_0/n_s. In typical situations this factor ranges in value from 10 to 10^6. As in the case of homogeneous nucleation of droplets from the vapor, n_s disappears† from the expression for equilibrium concentration of critical nuclei

$$n^\star = n_0 \exp (-\Delta G^\star/kT). \qquad \text{(C.16)}$$

It is evident that there is no rotational contribution to the free energy of formation of an embryo on a substrate for this case of localized adsorption of all cluster sizes. Also there is no problem in connection with conservation of degrees of freedom in this case so the second term of equation (B-17) for the free energy of "separation" is zero. Nevertheless the first term still obtains and thus the contribution from the free energy of separation becomes

$$\Delta G_s = kT \ln (2\pi i)/2. \qquad \text{(C-17)}$$

However, in view of the small values of i usually encountered, this term may be neglected and therefore the true free energy of formation of a critical nucleus may be represented by

$$\Delta G^{\star\prime} = \Delta G^\star + \Delta G_q. \qquad \text{(C-18)}$$

Applying this statistical correction to equation (C-11) one obtains†

$$J = C_1 p \exp [(\Delta G^\star_{\text{des}} - \Delta G^\star_{\text{sd}} - \Delta G^\star)/kT] \qquad \text{(C-19)}$$

* This treatment of the statistical correction follows the original suggestion of Lothe and Pound.[65] A treatment utilizing a different pedagogical approach but yielding the same result is included in an article by Hirth[87] which is based on Lothe and Pound's[65] work and on the statistical developments of Fowler and Guggenheim.[143]

† ΔG_g has been factored from the exponential.

in which*

$$C_1 = (\Delta G^{\star\prime}/4\pi kTi^{\star 2})^{\frac{1}{2}} \cdot 2\pi r^{\star}\, a \sin\theta n_0/(2\pi mkT)^{\frac{1}{2}}$$

whereupon equation (C-12) becomes

$$(1/\Delta G_{v_{\text{crit}}})^2 = \left[\frac{3}{16\pi\sigma_{c-v}{}^3\phi_3(\theta)}\right][kT(\ln C_1 + \ln p) + \Delta G^{\star}_{\text{des}} - \Delta G^{\star}_{\text{sd}}] \tag{C-20}$$

Accordingly a plot of $(1/\Delta G_{v_{\text{crit}}})^2$ versus $kT(\ln C_1 + \ln p)$ should give a straight line of slope $3/16\pi\sigma_{c-v}{}^3\phi_3(\theta)$ and intercept $3(\Delta G^{\star}_{\text{des}} - \Delta G^{\star}_{\text{sd}})/16\pi\sigma_{c-v}{}^3\phi_3(\theta)$. It will be shown in the part on Experiment that equation (C-20) provides a somewhat better description of data for the nucleation of crystals than equation (C-12).

One notes that there should be less of a problem relating to thermal accommodation upon addition of atoms to the developing embryo than in the case of nucleation in the vapor phase. The energy difference between the adsorbed and crystalline states is less than that between the vapor and crystalline states. Also the solid substrate should serve as an effective thermal sink for this heat of "condensation". Finally, unlike the situation in cloud-chamber work, the period of these experiments is sufficiently great that it is permissible to assume a steady-state distribution of embryos.

A second case arises if condition (C-8) does not hold and the embryos behave as a two-dimensional gas, as is possible for the nucleation of liquids or of solids only weakly bound to the substrate. In this situation, if the critical nuclei are moving more or less freely in the two-dimensional gas, rotational and translational degrees of freedom are activated. Thus instead of equations (C-6) or (C-18) one gets, analogously to equation (B-21),

$$\Delta G^{\star\prime} = \Delta G^{\star} + \Delta G_t + \Delta G_r' + \Delta G_s' \tag{C-21}$$

where

$$\Delta G_t = -kT \ln [(2\pi m_i kT)/n_s h^2], \tag{C-22}$$

$$\Delta G_r = -kT \ln [(8\pi^3 IkT)^{\frac{1}{2}}/h], \tag{C-23}$$

and I is the moment of inertia for one rotational degree of freedom.

Applying these corrections and the frequency factor equation (C-9) one obtains, in lieu of equation (C-19)

$$J = C_1'' p \exp [(\Delta G^{\star}_{\text{des}} - \Delta G^{\star})/kT] \tag{C-24}$$

in which

$$C_1'' = (\Delta G^{\star\prime}/4\pi kTi^{\star 2})^{\frac{1}{2}} 4\pi^2 r^{\star} \sin\theta i^{\star} kT(2\pi IkT)^{\frac{1}{2}}/h^3 \nu. \tag{C-25}$$

* For a given system, C_1 is approximately constant over a range of p and T and characteristically assumes a value of about 10^{17} dyne^{-1} sec^{-1}.

Accordingly in such a case equation (C-20) would be replaced by

$$(1/\Delta G_{v_{\text{crit}}})^2 = [3/16\pi\sigma_{c-v}{}^3\phi_3(\theta)][kT(\ln C_1'' + \ln p) + \Delta G^\star_{\text{des}}].^* \quad \text{(C-26)}$$

Again the test of equation (C-26) would involve a plot of $(1/\Delta G_{v_{\text{crit}}})^2$ versus $kT(\ln C_1'' + \ln p)$.

As discussed in Section B, there is some question in regard to the correctness of assigning macroscopic thermodynamic properties to a small aggregate of atoms such as a critical nucleus. Hence one should exercise caution in identifying the values of θ and σ_{c-v} obtained from critical supersaturation data by means of equation (C-20) or (C-26) with the thermodynamic quantities to be observed on bulk phases.

It should be possible to apply the thermodynamics of inhomogeneous systems,[(81–83)] as discussed in the development of equations (B-15) and (B-16) to this case of heterogeneous nucleation. Again, however, one notes that the cardinal assumption of low density gradient might not be correct at the crystal–vapor interface.

2. *Time-Dependent Adatom Concentration*

In cases where the adatoms are strongly bound to the substrate surface, e.g. silver on tungsten, $\Delta G^\star_{\text{des}}$ is large and thus the residence time τ_s is very long at low temperatures. Accordingly the relaxation period for equalization of re-evaporation and impingent fluxes is long with respect to the observation time in a typical experiment.† Nevertheless, as noted by Moazed and Pound,[(145)] one may assume metastable equilibrium between embryos and adatoms. However, the adatom population will no longer be given by equation (C-1) but will equal n_s' the integral condensate in time t,

$$n_s' = tp/(2\pi mkT)^{\frac{1}{2}}. \quad \text{(C-27)}$$

According to equation (C-1), the adatom population which would be in equilibrium with the vapor of the bulk condensate is $n_{s_e} = [p_e/2\pi mkT)^{\frac{1}{2}}]/\bar{\omega}\exp(-\Delta G^\star_{\text{des}}/kT)$. It is evident that in this case the supersaturation ratio is

$$(n_s'/n_{s_e}) < (n_s/n_{s_e}) = (p/p_e) \quad \text{(C-28)}$$

and

$$\Delta G_v' = -(kT/\Omega)\ln(n_s'/n_{s_e}). \quad \text{(C-29)}$$

Accordingly equation (C-19) for the nucleation rate becomes

$$J = C_2 n_s' \exp[-(\Delta G^\star_{\text{sd}} + \Delta G^\star)/kT] \quad \text{(C-30)}$$

where

$$C_2 = (\Delta G^{\star\prime}/4\pi kTi^{\star 2})^{\frac{1}{2}} 2\pi r^\star a \sin\theta\bar{\omega}n_0 \quad \text{(C-31)}$$

and, of course, $\Delta G_v'$ of equation (C-29) replaces ΔG_v.

* Assuming complete thermal accommodation.

† In fact it would seem physically impossible for the re-evaporation flux ever to equal the impingent flux in some cases.

If the binding energy of an adatom to the substrate is small (see Part A) α_c may be less than unity and, again, equation (C-1) will not yield the adatom concentration. In this case n_s'' the adatom concentration is the integral of the net sticking flux in time t. Such a problem is not amenable to solution in general but may be solved in specific cases. When this effect is important, the supersaturation ratio is given by

$$(n_s''/n_{s_e}) < (n_s/n_{s_e}). \tag{C-32}$$

3. *Lattice Strain and Interface Structure*

Turnbull and Vonnegut[60,146] have considered the possibility that the condensate-substrate interface is coherent* and hence that the nucleus is elastically strained. However, in most cases nucleation occurs either on high-index planes containing numerous monatomic ledges or in the vicinity of lattice imperfections such as grain boundaries or dislocations. In these situations it would seem unlikely that any degree of coherency exists between condensate and substrate.† Further, it appears reasonable to assume that the nucleus–vapor interfacial free energy is isotropic. Indeed, this is tacitly assumed in invoking the spherical cap model for the nucleus. In such a case it is plausible that the nucleus–substrate interfacial free energy could also be relatively insensitive to orientation of the interface, implying no appreciable effects of disregistry.

However, if coherent or semi-coherent nucleation occurs, the minimum energy configuration for a critical nucleus may be determined[60] by minimizing the sum of the strain energy of the nucleus and the interfacial free energy of the semicoherent interface σ_{c-x}.[147,148] The strain energy of the nucleus is assumed‡ to equal a constant times the square of the bilateral strain, i.e. to C_3e^2. The specific interfacial free energy may be represented by

$$\sigma'_{c-x} = \sigma^0_{c-x} + C_4(\delta' - e) \tag{C-33}$$

where C_4 is a constant, σ^0_{c-x} is the term due to bond type§ and $\delta' = \Delta a/a$ is the relative disregistry of the lattice spacings of the nucleus and substrate interface planes. Accordingly one employs σ'_{c-x} instead of σ_{c-x} and $(\Delta G_v + C_3e^2)$ in place of ΔG_v to obtain as the analog of equation (C-6)

$$\Delta G^\star = [16\pi\sigma_{c-v}{}^3\phi_3(\theta')/3(\Delta G_v + C_3e^2)^2] \tag{C-34}$$

* A coherent interface is defined as an interface across which there is perfect one to one atomic matching. Usually such planes are low-index planes for both phases meeting at the interface. Semi-coherent interfaces are interfaces which have one to one matching over most of the surface of contact but which contain some interface dislocations.[147,148]

† Coherency might be more likely in disc nucleation than in the present case of cap nucleation.

‡ This is only an approximation in the present case of cap nuclei where the bilateral strain would be expected to vary from e at the interface to zero at the top of the cap.

§ The chemical term may be dependent on surface orientation. For example, Fletcher[149] has pointed out that a polar substrate may orient H_2O dipoles thus reducing the entropy and increasing the free energy of formation of ice embryos.

where θ' refers to the value of the contact angle with $\sigma_{c-x} = \sigma'_{c-x}$ in equation (C-4). δ' is the upper limit on e and thus if $|\Delta G_v|_{\text{crit}} \gg C_3\delta'^2$ the nucleus will form coherently such that $e \simeq \delta'$ and $\sigma'_{c-x} \simeq \sigma^0_{c-x}$. On the other hand if $C_3\delta'^2 \gg |\Delta G_v|_{\text{crit}}$ the nucleus will form incoherently with $e \simeq 0$ and $\sigma'_{c-x} = \sigma^0_{c-x} + C_4\delta'$. For intermediate cases, $\Delta G^\star$ in equation (C-34) must be minimized with respect to e. An approximate result for such a case is given by Turnbull and Vonnegut.[60,146]

As noted above, the orientation of the substrate affects the nucleation process. If the substrate has an anisotropic surface free energy, σ_{x-v} for a low-index substrate plane should be less than that for a high-index substrate plane. Thus it is likely that $(\sigma_{x-v} - \sigma_{c-x})$ is larger for the high-index plane. Hence according to equations (C-4) and (C-6), the high-index substrate plane should be a more effective heterogeneous nucleation catalyst.

4. *Surface Imperfections on the Substrate*

A monatomic ledge separating two low-index facets of the substrate should be a more effective catalyst than the facets. To first order, the condensate-vapor surface area and the contact angle θ will be unaffected if a cap-shaped nucleus forms on a monatomic ledge as depicted in Fig. 12. In such a case

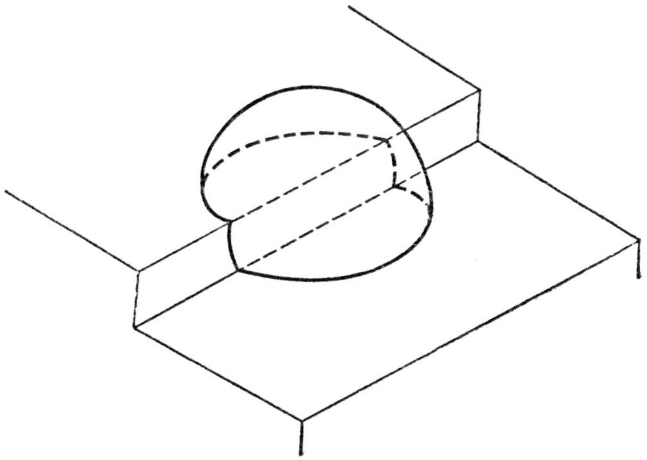

FIG. 12. Cap-shaped embryo forming on a monatomic ledge separating two low-index regions.

the free energy of formation of an embryo containing i atoms, equation (C-5) will be reduced by an amount

$$\begin{aligned}\delta\Delta G^0 &= 2r(\varepsilon_{x-v} - \varepsilon_{x-c}) \sin\theta \\ &= 2hr(\sigma_{x-v} - \sigma_{x-c}) \sin\theta\end{aligned} \qquad \text{(C-35)}$$

where the ε's are the edge energies of the monatomic ledge and h the step height. Thus because the free energy of formation of a nucleus is less when it forms on a monatomic ledge, the ledge provides a more effective catalytic site for nucleation.

Turnbull,[150] in considering the liquid–solid transformation, showed that solid retained in cylindrical microcavities in a substrate upon superheating a melt provides nuclei for transformation on subsequent supercooling. This is a specific application of the general idea that a hole or crevice in a substrate provides more area per unit volume of embryo or nucleus at which substrate–vapor interface is replaced by substrate–condensate interface. Hence for certain values of the relevant specific interfacial free energies σ, the free energy of formation of a nucleus at such a re-entrant site will be appreciably less than on a flat substrate and thus the hole or crevice will be a preferred site for nucleation.

Chakraverty and Pound[151] have developed the detailed kinetics for the case where the re-entrant site is a crevice formed by two planes of a macroscopic step meeting at an angle η as shown in Fig. 13. The analytical ex-

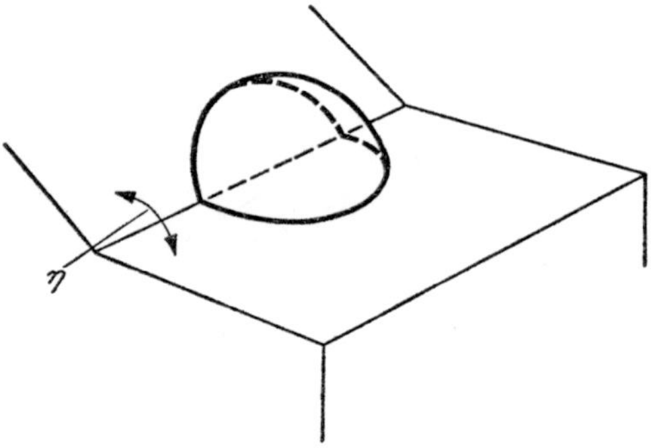

FIG. 13. Cap-shaped nucleus formed at a re-entrant angle η between two planes.

pressions are available only for the case in which $\eta = \pi/2$. The free energy of formation of the critical nucleus at the macroscopic ledge for which $\eta = \pi/2$ is

$$\Delta G_L^\star = (16\pi\sigma_{c-v}^3/3\Delta G_v^2)\phi_4(\theta) \qquad \text{(C-36)}$$

in which the complicated function of contact angle

$$\phi_4(\theta) = \tfrac{1}{4}\Big\{(\sin\theta - \cos\theta) + (2/\pi)\cos^2\theta\,(\sin^2\theta - \cos^2\theta)^{\frac{1}{2}} + (2/\pi)\cos\theta\sin^2\theta\sin^{-1}\cot\theta - \cos\theta\sin^2\theta - 2/\pi r \int_{r^\star\cos\theta}^{r^\star\sin\theta} \sin^{-1}[r^\star\cos\theta/(r^{\star 2} - y^2)^{\frac{1}{2}}]\,dy.\Big\} \qquad \text{(C-37)}$$

Here $r^\star = -2\sigma_{c-v}/\Delta G_v$ is the radius of curvature of the critical nucleus and y is a variable of integration. One notes that in the present case where the interfacial free energies are assumed to be isotropic the metastable equilibrium shape of the nucleus surface must be spherical.* Also the contact angle θ must be the same at all intersections of the nucleus surface with the substrate.* Further, for $\eta = \pi/2$ and $\theta \leq 45°$ any cluster at such a ledge is unstable with respect to growth at *any* supersaturation ratio greater than unity and $\Delta G_L^\star$ vanishes. The remainder of the expression for nucleation rate at a macroscopic ledge will be essentially the same as for nucleation on a flat surface (equation C-20 or C-30). Accordingly, the ratio of nucleation rate at the macroscopic ledge to that on a flat surface may be represented as

$$\ln (J_L/J) = (16\pi\sigma_{c-v}^3/3\Delta G_v^2)[\phi_3(\theta) - \phi_4(\theta)]. \qquad \text{(C-38)}$$

Under conditions of critical supersaturation, $(16\pi\sigma_{c-v}^3/3\Delta G_v^2)$ typically assumes values of the order of 28. A plot of equation (C-38) for $(16\pi\sigma_{c-v}^3/3\Delta G_v^2) = 28$ is given in Fig. 14 where it is seen that J_L is greater than J for all

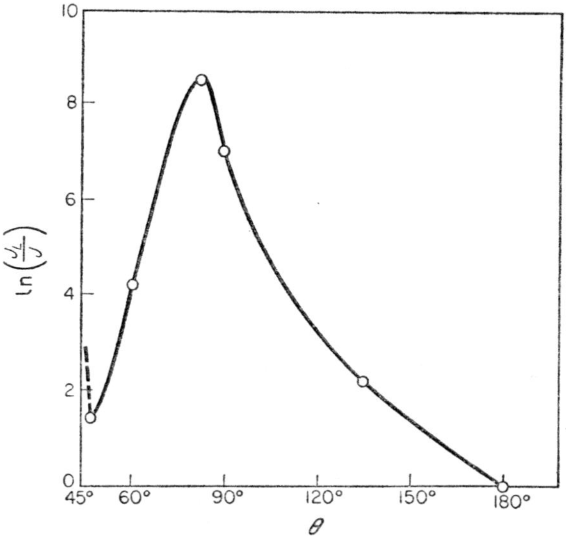

FIG. 14. Logarithm of the ratio of nucleation rate on a macroscopic ledge of $\eta = 90°$ to that on a flat surface as a function of contact angle from equation (C-38) with $(16\pi\sigma_{c-v}^3/3\Delta G_v^2) = 28$.

contact angles. From this plot, it would seem that appreciable decoration of macroscopic ledges should be expected only in systems where the contact angle lies between 60 and 120°.†

Also, if imperfections of varying catalytic potency are present on a surface, a situation may result wherein the nucleation rate at the most potent catalyst is rapid while the nucleation rate at other sites is negligible. If the spacing between the potent catalytic sites is large compared to the root mean square

* Ignoring line tension.
† Or is less than 45°.

path $\bar{X}$ (equation A-10), molecules adsorbing on the substrate within a distance $\bar{X}$ of the potent sites will diffuse to the nuclei growing there and condense, while other molecules will re-evaporate before reaching a growing nucleus. This will lead to a situation in which, even though nucleation has occurred, the sticking coefficient β, which is equal to the amount of material on the substrate after termination of the experiment divided by the integral amount that impinged there according to equation (A-2), may be less than unity. On the other hand if all sites on a surface have equivalent catalytic potency, one would expect from equation (C-12) or (C-20) that β should be unity at supersaturations greater than $(p/p_e)_{\text{crit}}$.

5. *Impurity Effects*

Impurity adsorption may affect the nucleation in several ways. Firstly, as discussed in Part A, impurity adsorption on the substrate is likely to decrease the binding energy of an adsorbed atom to the substrate and hence lower $\Delta G^{\star}_{\text{des}}$. This would decrease n_s according to equation (C-1) and J

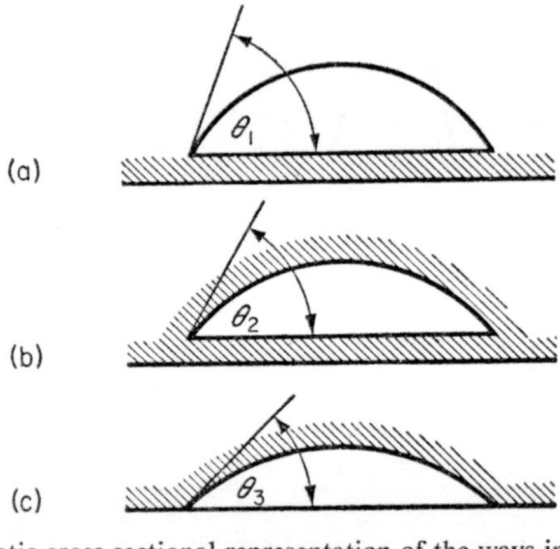

FIG. 15. Schematic cross-sectional representation of the ways in which impurity can affect nucleation. The hatching represents impurity adsorbate.

according to equation (C-19). Also, as discussed in Part A, a decrease in $\Delta G^{\star}_{\text{des}}$ increases the likelihood of reflection of an incident atom, leading to $\alpha_c < 1$. This effect would tend to increase the relaxation time for establishment of the steady-state adatom population n_s.

Finally, the adsorbate would affect $\Delta G^{\star}$ (equation C-6) and hence J (equation C-19) by changing $\phi_3(\theta)$. Equation (C-4) may be rewritten in the form

$$\cos\theta = (\sigma_{x-v} - \sigma_{x-c})/\sigma_{c-v}, \tag{C-39}$$

which shows that for a barrier to nucleation to exist $\sigma_{x-v} - \sigma_{x-c}$ must be less than σ_{c-v}. As $\cos\theta$ approaches unity, θ and $\phi_3(\theta)$ approach zero and the nucleation rate becomes very large. Now consider Fig. 15 which depicts several possible modes of impurity adsorption on a nucleus. Impurity adsorption decreases the specific surface free energies σ_{x-v} and σ_{c-v} because this is essentially equilibrium adsorption. However, the adsorption indicated at the crystal–substrate interface may not be equilibrium adsorption. In this situation σ_{x-c} may be either increased or decreased; it is likely that σ_{x-c} will be increased under the usual experimental conditions. Hence in case number 1 of Fig. 15, σ_{x-v} decreases, σ_{x-c} increases and cos θ and J decrease. In case number 2 σ_{c-v} and σ_{x-v} decrease while σ_{x-c} increases. If the relative decrease in σ_{c-v} is greater than the relative decrease in $(\sigma_{x-v} - \sigma_{x-c})$, J will increase, and vice versa. In case number 3 σ_{c-v} and σ_{x-v} decrease and σ_{x-c} is unchanged.* Hence as in case 2, J may increase or decrease depending on the relative decreases in σ_{c-v} and $(\sigma_{x-v} - \sigma_{x-c})$. However, the decrease in $(\sigma_{x-v} - \sigma_{x-c})$ will be less than in case 2 because σ_{x-c} is unchanged. Accordingly, case 3 should always yield a higher J than case 2. To summarize, in case 1 J should always be decreased by adsorption; in cases 2 and 3 the decrease in $\Delta G^{\star}_{\text{des}}$ will favor a decrease in J, but the change in $\phi_3(\theta)$ may favor either a decrease or an increase in J. On the whole it is likely that impurity adsorption will decrease J.

6. *Low Substrate Temperatures*

It has already been noted above that, at temperatures where τ_s becomes long compared with the experimental observation time, the nucleation kinetics are modified in that the adatom concentration becomes time dependent. At lower temperatures† the mean period for atomic-jump processes on the surface, $\tau_v = 1/\bar{\omega} \exp(-\Delta G^{\star}_{\text{sd}}/kT)$, may be large compared with the time required for a monolayer to accumulate on the substrate. In such a case the model which envisages embryos in equilibrium with single atoms breaks down. If the single incident atoms simply stick to the substrate at their impingement sites, an amorphous condensate could form in lieu of the crystalline condensate expected from the embryo-nucleus equilibrium model. This amorphous film may then crystallize by processes of nucleation and growth.

7. *Nucleation of a Metastable Phase*

A metastable phase may nucleate instead of the phase which is most stable at the substrate temperature. Although the thermodynamic driving force ΔG_v is less for nucleation of a metastable phase, its σ_{c-v} and/or σ_{x-c} may be sufficiently low to give a smaller $\Delta G^{\star}$. In fact this consideration is the theoretical basis of Ostwald's empirical "law of stages".[152]

* This case represents equilibrium adsorption (or lack of it) at the crystal–substrate interface.

† Lower because it is expected that $\Delta G^{\star}_{\text{sd}} < \Delta G^{\star}_{\text{des}}$.

8. *High Substrate Temperatures*

Of course at high substrate temperatures it is possible that atoms incident on a substrate may (a) diffuse into the bulk material by volume, grain-boundary or dislocation-pipe diffusion, (b) surface diffuse to other substrate regions or parts of the apparatus, or (c) chemically interact with the substrate or with impurity adsorbate. In any of these cases the effective surface supersaturation would be reduced and the nucleation kinetics modified.

9. *Sticking Coefficient, β*

Nucleation results are frequently reported in terms of a sticking coefficient β which is equal to the amount of material on the substrate after termination of the experiment divided by the integral amount that impinged there according to equation (A-2). Hence a value of $\beta = 1$ indicates that $(p/p_e) > (p/p_e)_{\text{crit}}$ while a value of $\beta = 0$ shows that $(p/p_e) \ll (p/p_e)_{\text{crit}}$. An intermediate value of β may be due to a situation in which $\alpha_c < 1$ (see section E). However, if $\alpha_c = 1$ in this intermediate example it indicates that (p/p_e) is only slightly less than $(p/p_e)_{\text{crit}}$. In this case some nuclei form but the density of these nuclei is so small that many adatoms re-evaporate before they can diffuse to and join a growing nucleus. Thus, while certainly not a sensitive test of the nucleation rate equations, β values provide some information about $(p/p_e)_{\text{crit}}$.

(b) *Heterogeneous Nucleation of a Disc-Shaped Nucleus on a Clean Surface*

1. *Constant Population of Adatoms*

When the interfacial free energy σ_{c-v} is anisotropic such that low-index surfaces are present in the Gibbs–Wulff shape, the configuration of minimum free energy for a nucleus may be a disc on which the circular surface is a low-index plane as illustrated in Fig. 16a. In this case the free energy of formation of an embryo is

$$\Delta G^0 = 2\pi r \varepsilon_{c-v} + \pi r^2(\sigma_{c-v} + \sigma_{c-x} - \sigma_{x-v}) + \pi r^2 h \Delta G_v. \qquad \text{(C-40)}$$

Maximizing ΔG^0 with respect to r, one obtains the free energy of formation of the critical nucleus* exclusive of the statistical effect[65]

$$\Delta G^\star = -\pi \varepsilon_{c-v}^2/h[\Delta G_v + (\sigma_{c-v} + \sigma_{c-x} - \sigma_{x-v})/h]. \qquad \text{(C-41)}$$

The frequency factor $\bar{\omega}$† and the non-equilibrium factor Z are almost[144] the same as in the previous treatment of a cap-shaped nucleus, except that the

* Note that in general the specific ledge free energy $\varepsilon_{c-v} \neq h\sigma_{c-v}$.

† In this case of crystal nucleation the two-dimensional gas model of equation (C-9) does not obtain.

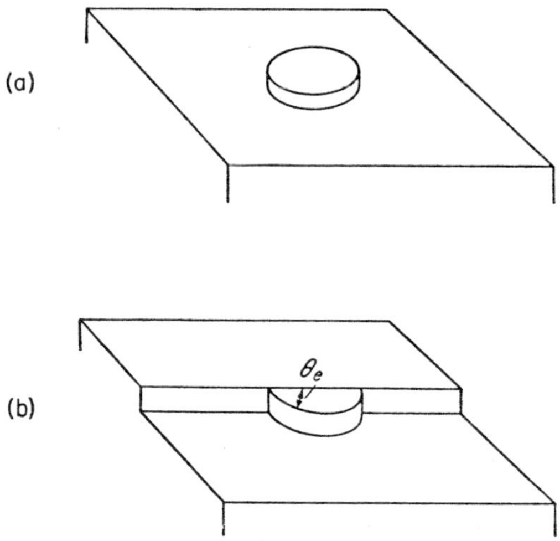

FIG. 16. (a) Disc-shaped embryo on a heterogeneous substrate. (b) Disc-shaped embryo at a monatomic ledge.

circumference is $2\pi r$ instead of $2\pi r \sin\theta$. Thus, substituting in equation (C-7) and reducing the resulting expression, one gets

$$\begin{aligned} J &= C_5' p^2 \exp\,[(2\Delta G^\star_{\text{des}} - \Delta G^\star_{\text{sd}} - \Delta G^\star)/kT], \\ C_5' &= (Zr^\star a/mkT\bar{\omega}). \end{aligned} \tag{C-42}$$

At the critical supersaturation where $\ln J = 0$,

$$\begin{aligned} &\Delta G_{v_{\text{crit}}} + (\sigma_{c-v} + \sigma_{c-x} - \sigma_{x-v})/h \\ &\quad = -\pi\varepsilon_{c-v}{}^2/h[kT(\ln C_5' + 2\ln p) + 2\Delta G^\star_{\text{des}} - \Delta G^\star_{\text{sd}}]. \end{aligned} \tag{C-43}$$

One notes that there is no simple plot by means of which one may determine the quantities appearing in equation (C-43). However, knowing ε_{c-v} and $\Delta G_{v_{\text{crit}}}$ for various values of $kT(\ln C_5' + 2\ln p)$, one can solve pairs of simultaneous equations to determine the quantities $(2\Delta G^\star_{\text{des}} - \Delta G^\star_{\text{sd}})$ and $(\sigma_{c-v} + \sigma_{c-x} - \sigma_{x-v})/h$. Also by measuring $\Delta G_{v_{\text{crit}}}$ at various values of p and T one may determine whether the nucleation kinetics follow equation (C-12) for cap-shaped nuclei or equation (C-43).*

In the usual case where the statistical effect (equation C-14) is important, i.e. where the number of sites n_0 on the substrate surface greatly exceeds the population n_s of adatoms, one must correct equation (C-42) by the factor n_0/n_s to yield

$$\begin{aligned} J &= C_5 p \exp\,[(\Delta G^\star_{\text{des}} - \Delta G^\star_{\text{sd}} - \Delta G^\star)/kT], \\ C_5 &= Z \cdot 2\pi r^\star a n_0/(2\pi mkT)^{\frac{1}{2}}. \end{aligned} \tag{C-44}$$

* The results for the two models converge at small nucleus sizes and hence differentiation is difficult in this case.[170]

For the critical condition

$$\Delta G_{v_{\mathrm{crit}}} + (\sigma_{c-v} + \sigma_{c-x} - \sigma_{x-v})/h = -\pi\varepsilon^2_{c-v}/h[kT(\ln C_5 + \ln p) + \Delta G^\star_{\mathrm{des}} - \Delta G^\star_{\mathrm{sd}}] \qquad \text{(C.45)}$$

and as indicated above one can determine $(\sigma_{c-v} + \sigma_{c-x} - \sigma_{x-v})/h$ and $\Delta G^\star_{\mathrm{des}} - \Delta G^\star_{\mathrm{sd}}$ from experimental data. Again, by measuring $\Delta G_{v_{\mathrm{crit}}}$ at various values of p and T one may in principle determine whether the nucleation kinetics follow equation (C-20) for cap-shaped nuclei or equation (C-45).

2. *Time-Dependent Adatom Concentration*

The required modification of the kinetics in this situation is the same as for cap nucleation in that n_s', equation (C-27), replaces n_s, equation (C-1).

3. *Lattice Strain and Interface Structure*

The possibility of coherent or semicoherent nucleation seems somewhat greater for disc nucleation than for cap nucleation, especially if the substrate is a low-index plane. As before, in such a case $\sigma_{c-x} = \sigma_{c-x}{}^0 + C_4(\delta' - e)$ and

$$\Delta G^\star = -\pi\varepsilon^2_{c-v}/h\{\Delta G_v + C_4e^2 + [\sigma_{c-v} + \sigma^0_{c-x} + C_4(\delta' - e) - \sigma_{x-v}]/h\}. \qquad \text{(C-46)}$$

Again, the transition from coherent to incoherent nucleation will depend on the magnitude of ΔG_v relative to the rest of the terms in the denominator of equation (C-46).

4. *Substrate Perfection*

The nucleation of a disc may be catalyzed by either macroscopic or monatomic ledges as shown in Fig. 16b. For this case

$$\Delta G^0 = [(h\Delta G_v + \sigma_{c-v} + \sigma_{c-x} - \sigma_{x-v})r^2 + 2rh\varepsilon_{c-v}]\phi_5(\theta_e) \qquad \text{(C-47)}$$

where θ_e is the angle depicted in Fig. 16b and $\phi_5(\theta_e) = \theta_e - \sin\theta_e \cos\theta_e$. Accordingly equation (C-41) becomes

$$\Delta G^\star = -\varepsilon^2_{c-v}\phi_5(\theta_e)/h[\Delta G_v + (\sigma_{c-v} + \sigma_{c-x} - \sigma_{x-v})/h]. \qquad \text{(C-48)}$$

Thus nucleation at a monatomic ledge will be more rapid than on a flat surface by an amount determined by $\phi_5(\theta_e)$. Note the analogy to the enhancement of cap nucleation with respect to homogeneous nucleation as determined by $\phi_3(\theta)$.

5. *Impurity Effects*

Impurity adsorption would have effects similar to those considered for cap nucleation. Although the kinetics are not developed in detail it is noted that adsorption would lower $\Delta G^\star_{\mathrm{des}}$ as discussed above. Also, $\Delta G^\star$ would be

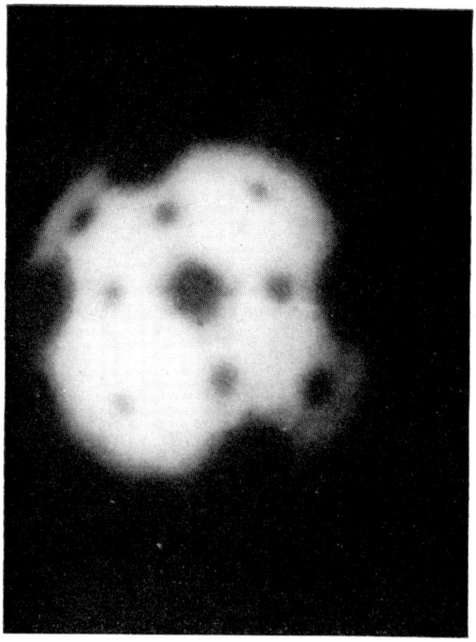

FIG. 17(a)

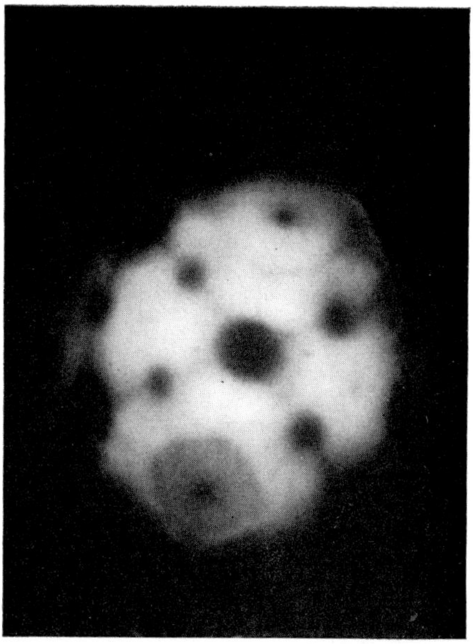

FIG. 17(b)

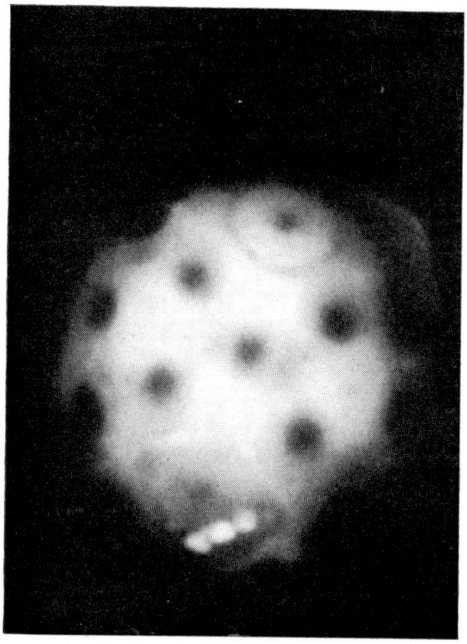

FIG. 17(c)

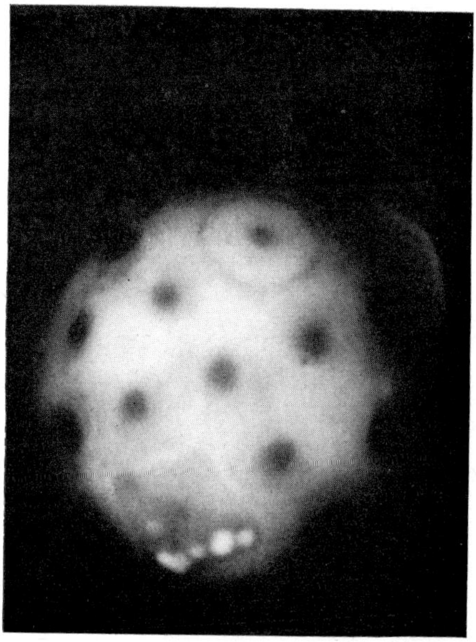

FIG. 17(d)

FIG. 17. Sequence of field-emission patterns in the course of nucleation of silver on a tungsten substrate, $10^6\times$ (reduced to 8/10ths), showing: (a) clean tungsten substrate, (b) adsorbed silver, (c–e) nucleation and growth of silver particles.

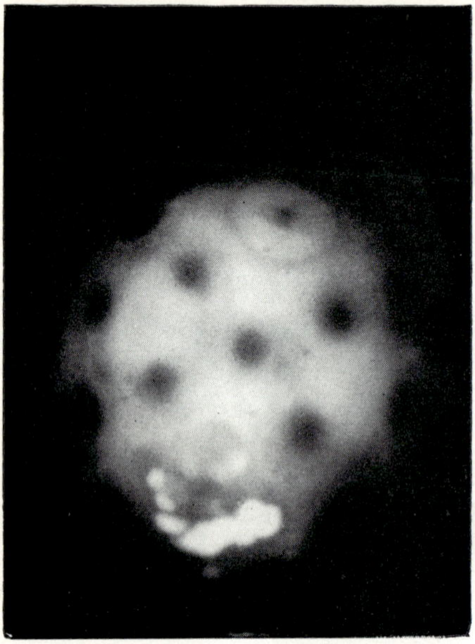

FIG. 17(e)

FIG. 17. Sequence of field-emission patterns in the course of nucleation of silver on a tungsten substrate, $10^6 \times$ (reduced to 8/10ths), showing: (a) clean tungsten substrate, (b) adsorbed silver, (c–e) nucleation and growth of silver particles.

changed by the alteration of the σ values. Both effects would influence J according to equation (C-44). Again, the most likely result of impurity adsorption will be to lower J.

(c) *Summary of Theory*

The existing theories of both homogeneous and heterogeneous nucleation suffer from a common deficiency: inherent in them is the tenuous assumption that macroscopic thermodynamic properties may be ascribed to small clusters of molecules. The ramifications of this assumption were discussed at some length in Section B on Nucleation in the Vapor Phase. Also there is a statistical mechanical contribution[65] (equation C-14) to the free energy of formation of the critical nucleus which is analogous to the translational term in homogeneous nucleation from the vapor. Two cases are distinguished in which rotational and translational degrees of freedom may or may not be activated in the formation of a critical nucleus. Accordingly equations (C-20), (C-26) and (C-45) are most useful in describing, analyzing and predicting phenomena involving heterogeneous nucleation from the vapor. Also equations (C-38) and (C-48) should prove useful in analyzing the effects of substrate imperfections, such as macroscopic steps, on the nucleation process.

It is possible that lattice disregistry between nucleus and substrate may affect the nucleation kinetics. However, any appreciable effect of this kind seems rather unlikely.

Finally, in view of the number of parameters involved, the problem of quantitative analysis of the effects of impurity adsorption on nucleation appears quite formidable.

3. Experiment

(a) *Measurements of Nucleation Rates*

1. *High-Vacuum Work*

In vacua of 10^{-6} mm Hg the residual-gas flux striking a substrate is equivalent to about one monolayer per second, which is roughly the same as the beam flux in typical nucleation experiments. Thus even when substrates have been carefully outgassed, it is not certain that contaminants are absent unless the residual vacuum is $\leqslant 10^{-6}$ mm Hg.

Moazed and Pound[145] performed a nucleation experiment by means of field-emission microscopy in a vacuum of $\sim 10^{-10}$ mm Hg. They observed a monocrystalline tungsten emitter tip while it was being subjected to impingement from a thermal beam of silver atoms. It was possible to distinguish field-emission patterns corresponding to: (a) a clean tungsten tip, (b) a tip contaminated with impurity, (c) a tip partially covered with adsorbed silver atoms and (d) silver nuclei on the tungsten tip. Because the field-emission microscope has a resolution of ~20 Å,(153) they were able to observe the nucleation process from its inception.* It was found that the nucleation occurred preferentially on the high-index planes. Figure 17 shows a sequence of stages in the nucleation process. In this experiment the mean residence time τ_s was sufficiently large that the nucleation followed equation (C-30). At a substrate temperature of 300°K, the critical value of n_s' for nucleation was 1.7×10^{14} cm^{-2} or about 0.2 of a monolayer, corresponding to $\Delta G'_{v_{\rm crit}} \simeq -2900$ cal/cc and $\theta \simeq 84$ degrees. Also, the critical value of n_s' was the same for fluxes of either 6×10^{11} or 1.4×10^{12} cm^{-2} sec^{-1}, indicating that indeed metastable equilibrium existed between embryos and single adatoms. This latter finding is most significant in that it is the first direct evidence that the embryo-adatom equilibrium model obtains in nucleation.†

Although the macroscopic contact angle of silver on tungsten has not yet been reported in the literature, it is thought(154) to be close to 0°.‡ Thus θ should be interpreted here merely as a parameter which indicates the magnitude of some impediment to the nucleation process. In the present case of nucleation of metal crystals on a clean metal substrate, it is possible that most of the adsorbed species are positive adions rather than neutral adatoms. If this is so it seems reasonable to suppose that mutual electrostatic repulsion precludes participation of the adions in the nucleation process and thus the effective concentration of adsorbed species is much lower than the total. As discussed by Becker(157) and Dekker,(158) the ratio of the equilibrium concentration of adions to adatoms is given by

$$n_{s_{\rm ion}}/n_s = \exp\,(E_{\rm adion} - E_{\rm adatom} - E_{\rm ion} + E_W)/kT \qquad \text{(C-49)}$$

in which partition functions are neglected and $E_{\rm adion}$ and $E_{\rm adatom}$ are the potential energies of desorption for ions and atoms, respectively. $E_{\rm ion}$ is the ionization potential of the condensate atoms, and E_W is the work function of the substrate. This ratio may be large in the silver–tungsten system. Although $E_{\rm adatom}$ and $E_{\rm adion}$ are not known, one can predict that large values of $(E_{\rm ion} - E_W)$ will favor atomic instead of ionic adsorption and thus lessen the impediment to nucleation. It would be interesting to repeat this nucleation experiment with another pair§ of metals for which $(E_{\rm ion} - E_W)$ is greater to

* Of course, the embryos and critical nuclei could not be resolved.

† Gretz and Pound(528) have recently confirmed these results for silver on tungsten and extended the study to include the temperature dependence of the critical supersaturation ratio for nucleation of cadmium on tungsten field-emitter tips. Their findings are in accord with the recommended theoretical relationships. More significantly they found that, although the critical supersaturation ratio for cadmium deposition is only about 65 (in contrast to ~10^{39} for silver), the critical adatom population is again of the order of one-tenth of a monolayer.

‡ There is still some controversy on this point(155,156) since it appears that θ may be quite sensitive to impurity adsorption in this case.

§ For Au–Mo, $(E_{\rm ion} - E_W)$ is 5 eV, versus only 3 for Ag–W.

see if the nucleation is indeed enhanced. Of course, it should be instructive to conduct all of these experiments over a wide range of substrate temperatures.

Chirigos *et al.*[160] studied the nucleation of silver on polycrystalline copper in both high (10^{-10} mm Hg) and low (10^{-5} mm Hg) vacua and illustrated the importance of the effect of vacuum on nucleation measurements. They projected a silver beam which contained a radioactive tracer on a copper substrate and monitored the amount of condensed silver with a Geiger counter. Their results are shown in Table 4.

TABLE 4

STICKING COEFFICIENTS β FOR Ag ON Cu

Target temperature, °C	Source temperature, °C	Beam flux, no./cm² sec	Exposure, hours	Residual gas pressure, mm Hg	p/p_e	β
345	536	4×10^8	48	$<4 \times 10^{-9}$	$\sim$ 460	1.0
320	540	4×10^8	72	$\sim 10^{-5}$	$\sim$1200	0.51
202	556	1×10^9	48	$\sim 10^{-5}$	$\sim 10^{10}$	0.66
400	740	2×10^{11}	1–9	$\sim 10^{-6}$	$\sim$4700	0.45–0.60

The values of $\beta < 1$ obtained in the experiments at poor vacua were time-dependent and increased with time of exposure. For example, β increased from 0.45 after 1 hour to 0.60 after 9 hours in the experiment at 400°C. Thus nucleation must have been occurring but at much lower rates than in the experiments at high vacua, even though the supersaturations were much greater. This experiment indicates the effect of impurity contamination in reducing nucleation rate as discussed in Section C-2-a-(5).

Ptushinskii[161] studied the nucleation of silver on an outgassed polycrystalline molybdenum substrate in a vacuum of 3×10^{-9} mm Hg at high supersaturations ($T_{source} = 973 - 1113$°K, $T_{substrate} = 300 - 800$°K). The measured β was unity under these conditions* while in vacua of 10^{-4} to 10^{-6} mm Hg it was 0.9. For silver ($T_{source} = 1090$°K) incident on glass, mica and single crystals of germanium at 300°K, he found $\beta = 0.89$, 0.92 and 0.96, respectively. In these cases the vacuum was $\sim 10^{-8}$ mm Hg and the system was not as thoroughly outgassed. Thus these results might be explained by impurity adsorption. Alternatively, one might argue that they are due to the effect of lattice disregistry (equation C-33).

Mayer and Gohre[163] recently studied the deposition of mercury onto quartz-glass substrates in vacua of about 10^{-10} mm Hg. They continuously

* Similarly, Von Goeler and Luscher,[162] by means of a radioactive tracer technique, found a sticking coefficient β of unity for gold on polycrystalline molybdenum substrates. The residual gas pressure was 2×10^{-9} mm Hg and unspecified ranges of substrate and beam temperatures were employed.

measured the condensation rate by means of a microbalance. At a beam flux of 6.5×10^{13} atoms/cm² sec, they found that β was unity at a substrate temperature of -133°C. However, at higher substrate temperatures (lower supersaturations) β increased monotonically from a low initial value as shown in Fig. 18. At a higher beam flux of 5.2×10^{14} atoms/cm² sec and the same substrate temperature of -66°C they found that β increased more rapidly than for the lower flux. Further, the critical supersaturation ratio

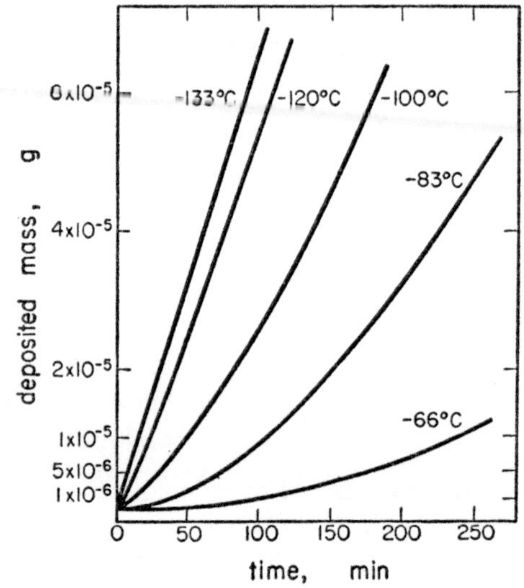

FIG. 18. Mercury condensation on quartz glass as a function of time in vacua of about 10^{-10} mm Hg and with a beam flux of 6.5×10^{13} atoms/cm² sec.(163) The -133°C curve corresponds to complete condensation ($\beta = 1$).

(probably for nucleation on imperfections, see discussion after equation C-38) at 6.5×10^{13} atoms/cm² sec and a substrate temperature of about -51°C was $(p/p_e)_{crit} \cong 10$. They noted that the low critical supersaturation ratio in this case of presumed low binding energy and the finding that $\beta = 1$ at a substrate temperature of -133°C indicate that the "hot atom" hypothesis of Sears and Cahn(47) did not apply. Their results may be interpreted to show agreement with the model of equations (C-12) and (C-20). At the lowest substrate temperature the mean free path for surface diffusion is large (see equations A-10 and D-38) and the spacing between nuclei is small because of the high nucleation rate. One observes that this high nucleation rate is due to the high supersaturation resulting from the low substrate temperature. Thus all atoms incident on the substrate adsorb and can diffuse to a growing nucleus. At higher substrate temperatures the mean free path for surface diffusion is less (see equations A-10 and D-38), and the nucleation rate is lower (lower supersaturation) so that initially some atoms may desorb before diffusing to a growing nucleus. However, as time goes on more nuclei form until eventually β again equals unity. These latter cases in which β is initially small and monotonically increases towards a limiting value of unity may indicate that catalytic sites of varying potency are present on the quartz-glass substrate as considered in the discussion following equation (C-38). This interpretation suggests that the sites of high catalytic potency exhibited a critical supersaturation ratio, from equation (C-38), of $\sim$10; on the other hand, general nucleation on the substrate occurred (for $\beta = 1$) at a critical supersaturation ratio of about 10^{10}, in good agreement with equation (C-12) or (C-20).

2. *Low-Vacuum Work**

Knudsen(164) (using ammonium chloride, mercury, zinc, cadmium, magnesium, copper and silver) and Wood(21) (using mercury, cadmium and iodine) carried out experiments in which partial pressures of the various materials listed above were established over glass substrates containing temperature gradients. They found that a "critical" substrate temperature existed above which no condensation occurred even though the system was supersaturated. Chariton and Semmenoff(165) carried out similar experiments for cadmium deposition on paraffin, picein and mica. They also found a critical temperature for condensation but suggested that this critical temperature varied slightly with change in the atomic flux incident on the substrate.

Motivated by the above work, Frenkel(139) developed a theory for condensation which was the precursor of the more sophisticated theory outlined in Part C-2. The principal difference between the two theories is that Frenkel assumed the activated species to be adatom doublets rather than polyatomic clusters and ignored the statistical factor.

Estermann(166) then studied the condensation of an atomic beam of cadmium on polycrystalline silver and copper targets as a function of beam flux and temperature and demonstrated that indeed the critical condition for nucleation depended on $(p/p_e)_{crit}$, which of course (see equation C-20) depends on both beam flux and temperature. In this work Estermann also demonstrated that adatoms are mobile by inserting cooled collimator slits between the beam source and target. He noted that the area on which deposition occurred was always somewhat less than the actual exposed area.

Using a thermal beam technique, Cockcroft(167) measured the critical supersaturation for deposition of cadmium on polycrystalline copper over a range of p and T. His method was somewhat indirect. He projected a thermal beam onto a flat surface and increased its intensity to a value just necessary for deposition (appreciable nucleation rate). However, instead of reporting this as the critical flux and using it to compute $(p/p_e)_{crit}$, he considered only the flux at the periphery of a disc of condensed metal on the substrate when it had reached its maximum diameter under the conditions of the

* In all the work reported in this section, the residual vacuum was 10^{-4} to 10^{-6} mm Hg.

experiment. This latter flux has no significance in relation to the nucleation process; it is merely the flux required to replace the loss due to surface diffusion away from the periphery at steady state. Fortunately however, under the conditions of the experiment, this flux is insignificantly less (by a factor of the order of $\frac{1}{2}$) than the true critical beam flux. Pound *et al.*[142] showed that Cockcroft's data follows equation (C-12) or (C-20) as shown in Figs. 19 and 20. Cockcroft found that at a substrate temperature of 103°K $\Delta G_{v_{crit}}$ was

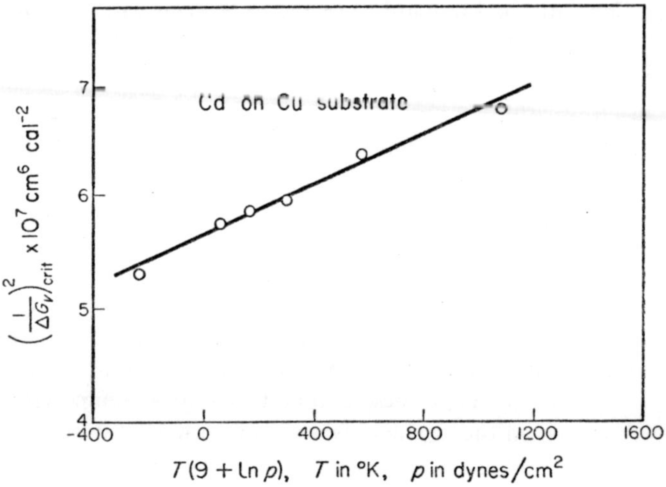

FIG. 19. Dependence of $\Delta G_{v_{crit}}$ on temperature for cadmium deposition on copper, equation (C-12).

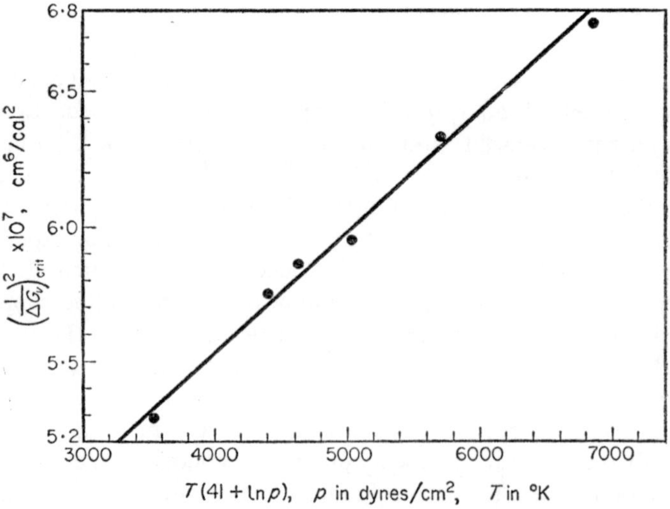

FIG. 20. Dependence of $\Delta G_{v_{crit}}$ on temperature for cadmium deposition on copper, equation (C-20).

identical for deposition on polycrystalline copper or silver, or glass. He interpreted this as nucleation on a common impurity adsorbate. Cockcroft repeated the Cd–Ag experiment on a substrate of freshly deposited polycrystalline silver and found that $(p/p_e)_{crit}$ was lower by a factor of 10. Thus it seems certain that impurity effects were present. Also it appears likely that θ, $\Delta G^{\star}_{des}$, and $\Delta G_{v_{crit}}$ refer to models 1 or 2 in Fig. 15.

Yang *et al.*[168] studied the nucleation of sodium from an atomic beam on polycrystalline silver, platinum, copper, nickel and cesium chloride over a

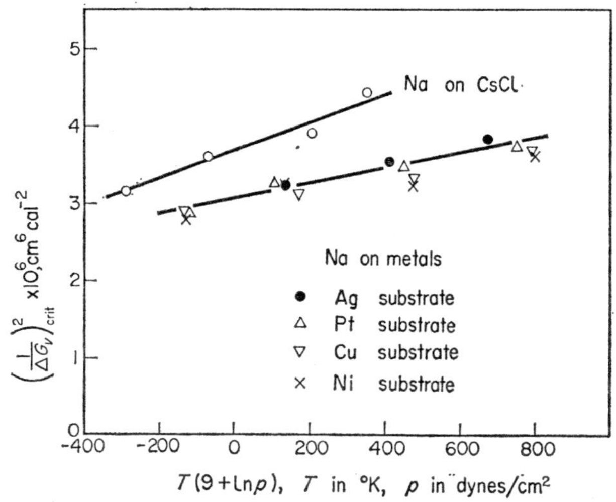

FIG. 21. Dependence of $\Delta G_{v_{crit}}$ on temperature for Na deposition on various substrates, equation (C-12).

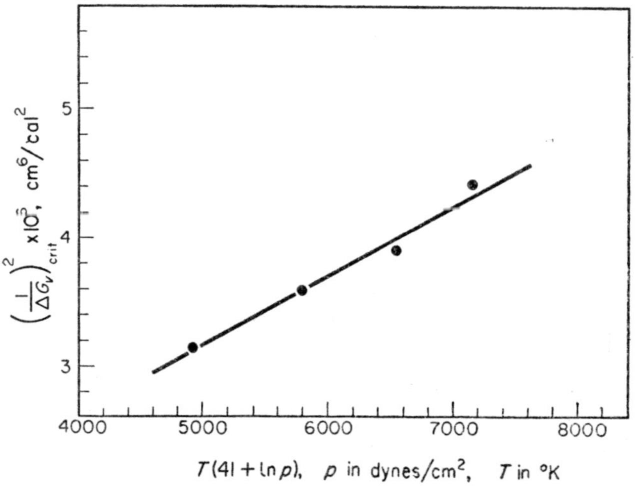

FIG. 22. Dependence of $\Delta G_{v_{crit}}$ on temperature for sodium deposition on CsCl, equation (C-20).

range of p and T. Their results were described by equation (C-12) as shown in Fig. 21 but a somewhat better fit is obtained with equation (C-20) as seen in Fig. 22.*[170] The deposition on the metal substrates was interpreted as nucleation on an impurity adsorbate; it was suggested that less impurity was present on the cesium chloride. The results of this experiment and Cockcroft's are listed in Tables 5 and 6.

TABLE 5

VALUES OF σ_{c-v}, θ, AND $\Delta G^{\star}_{des}$ FROM EQUATION (C-12)

Experiment	σ_{c-v} (ergs/cm^2)	Results	
		(θ degrees)	$\Delta G^{\star}_{des}$ (cal/g-atom)
Na on metals	350	101	5,750
Na on CsCl	350	81	4,220
Cd on Cu	760	99	10,300

TABLE 6

VALUES OF σ_{c-v}, θ, AND $\Delta G^{\star}_{des}$ FROM EQUATION (C-20)[170]

Experiment	σ_{c-v} (ergs/cm^2)	Results	
		θ (degrees)	$\Delta G^{\star}_{des}$ (cal/g-atom)
Na on CsCl	310	125	2,300
Cd on Cu	700	141	17,200

Sears and Cahn[47] have suggested that the large values of critical supersaturation noted in the above work are due to a situation in which the adatoms are appreciably hotter than the substrate, i.e. intermediate in temperature between source and substrate. As noted in Section A on Thermal Accommodation, this hypothesis seems unlikely from a theoretical point of view, i.e. it appears improbable that the relaxation time τ_r for thermal equilibration of an adatom with a substrate should be large relative to the mean period τ_s of residence on the surface. Moreover, in establishing the necessity of such a hypothesis they considered the case of nucleation of silver on silica and presumed[47] an observed critical[169] supersaturation ratio, $(p/p_e)_{crit}$, of 10 at a substrate temperature of 850°C. They extrapolated this number with

* One notes that the relatively high values of $\Delta G^{\star}_{des}$ preclude application of equation (C-24) for the two-dimensional gas model.

the aid of equation (C-42) to estimate the critical ratio at 192°C and obtained a value of 10^6, which is in contrast to the observation of Yang *et al.*[168,183] that it is about 10^{12} for nucleation of silver on glass at 192°C. However, Hruska* has pointed out that the observed critical supersaturation ratio of 10 at 850°C is for the cessation of silver whisker formation on a silica substrate rather than for nucleation of silver crystallites on the vitreous silica surface. Also Hruska has shown that a critical ratio of 100 at 850°C, as obtained by a reasonable theoretical estimate using equation (C-20), extrapolates to about 10^{12} at 192°C.† Thus it would appear that the necessity for the "hot adatom" hypothesis vanishes, at least for the present cases of high binding energy.

Rhodin[171] measured $(p/p_e)_{\text{crit}}$ as a function of p and T for aluminum condensation on sodium chloride, mica and glass. Inasmuch as the appropriate nucleation theory had not yet been developed, he reported his results as $\log p_{\text{crit}}$ vs. $1/T$. Straight lines were obtained‡ but the Knudsen beam flux was calculated from older vapor pressure data. Recalculation,§ using more recent vapor pressure data,[172] yields results which cannot be meaningfully interpreted on $\log p_{\text{crit}}$ vs. $1/T$ plots. Rhodin's observations are valuable because they indicate that the catalytic potency of these substrates increases in the order (100) Al//glass, (110) Al//NaCl, (110) Al//mica, (100) Al//NaCl, and (111) Al//mica.

Gunther[173] measured sticking coefficients as a function of beam flux for SiO and B_2O_3 on polycrystalline iron substrates in which a temperature gradient was maintained. From intercepts at $\beta = 0$ he found the critical beam flux for nucleation in the case of SiO. His data indicate that indeed the critical supersaturation for nucleation increases with decreasing substrate temperature, as one would predict from equation (C-20). However, the data for B_2O_3 indicate that plots of β vs. beam flux at constant substrate temperature are double-valued, a finding in complete disagreement with theory. He suggested that impurity adsorption may have affected his results. However, it seems likely that these nucleation processes may be complicated by reactions between adsorbed species on the substrate surface.

Yeh and Siegel[174] and Wells and Siegel[175] have studied the early stages of condensation of cadmium and silver, respectively, onto mica substrates. They deposited these metals for fixed times in a vacuum of $\sim 10^{-5}$ mm Hg and then counted the number of nuclei by electron microscopy of surface replicas. Their results are not measurements of absolute nucleation rates, but rather of rates that are influenced by nucleation, growth and agglomeration of crystallites. Nonetheless they found that the number of crystallites per unit area increased with decreasing substrate temperature in agreement

* Private communication.

† Indeed this $(p/p_e)_{\text{crit}}$ of 100 is consistent with Sears' description of his experiments.[169]

‡ Note from equation (C-20) that the slope of such a plot will be

$$\{(16\pi\sigma_{c-v}{}^3\Omega^2\phi_3(\theta)/3k^2)/[kT(\ln C_1 + \ln p) + \Delta G^{\star}_{\text{des}} - \Delta G^{\star}_{\text{sd}}]\}^{\frac{1}{2}},$$

which over a narrow temperature range will be $\sim$ constant.

§ S. J. Hruska, private communication.

with equation (C-19). Also, they noted that nucleation took place preferentially at defects such as ledges on the mica substrate in agreement with equation (C-35) or (C-38).

Fray and Nielsen(176) condensed zinc onto glass in a vacuum of $\sim 10^{-5}$ mm Hg and found that high supersaturations were required for nucleation on freshly melted glass which replica electron microscopy indicated was free from imperfections. On other glass substrates nucleation took place at lower supersaturations at imperfections such as pits and scratches, in agreement with equation (C-38). Even on these latter substrates, however, higher supersaturations were required if the glass was "poisoned" by hydrocarbons, indicating the effect of impurity adsorption.

Wegener(177) studied the nucleation and growth of potassium on a quartz substrate. Once nucleation occurred, the sticking coefficient was unity for $T_{\text{source}} = 60°\text{C}$ and $T_{\text{substrate}} = -180$, 0 or 21°C. At $T_{\text{source}} = 60°\text{C}$ the measured $(p/p_e)_{\text{crit}}$ was 1.008 for $T_{\text{substrate}} = 36°\text{C}$.

Devienne has used a radioactive-tracer technique to measure sticking coefficients at low beam fluxes (10^{-5} to 10^{-2} monolayers/sec) of antimony on glass, copper, aluminum and mica,(178,179) cadmium on copper, glass, aluminum and mica,(179) gold on copper, glass and aluminum(180) and silver on glass.(181)*† Typical of his results are those for antimony in Table 7.

TABLE 7

STICKING COEFFICIENT FOR ANTIMONY ON COPPER AND GLASS AS A FUNCTION OF AVERAGE "FILM" THICKNESS‡

Substrate T							
Copper 20°C	Å thick β	1.9 0.401	2.2 0.417	94.3 0.640	209 0.771	341 0.763	438 0.772
Copper 170°C	Å thick β	0.63 0.291	4.5 0.325	12.2 0.52	132 0.768		
Glass 20°C	Å thick β	1.3 0.311	2.6 0.307	31.7 0.464	132 0.768	497 0.829	
Glass 120°C	Å thick β	0.106 0.046	0.47 0.012	1.1 0.053	109 0.122		

In general, the exact beam fluxes were not reported. About all one can conclude from this work is that the nucleation rate was small at these low beam fluxes. Therefore the distance between nuclei was large and many adsorbed atoms re-evaporated before they could diffuse to a growing nucleus. As time

* This work is reviewed in reference 181.

† The metallic substrates were polycrystalline.

‡ The deposit undoubtedly consisted of many small crystallites distributed over the substrate surface.

went on, more nuclei formed and the re-evaporation probability decreased. At these low beam fluxes it is quite probable that impurity contamination decreased the nucleation rate. In fact, Garin and Prugne[182] found that when a glass substrate was cleaned by gas discharge, β approached unity for cadmium under experimental conditions similar to those for which Devienne[179] found $\beta \ll 1$.

Yang *et al.*[183] measured β for silver on polycrystalline gold and glass and obtained the results shown in Fig. 23. They used radioactive-tracer techniques

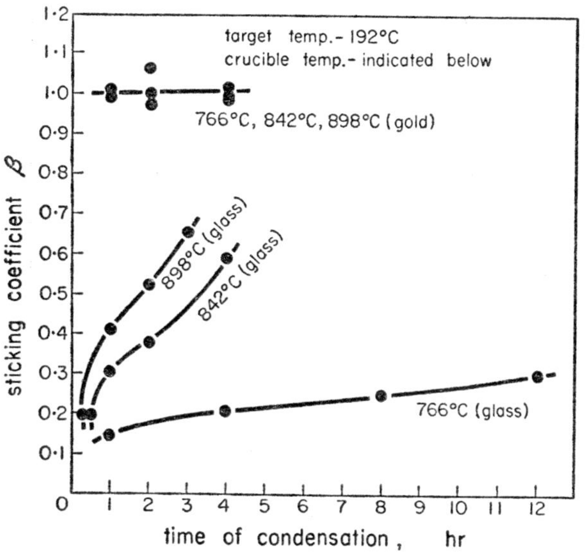

FIG. 23. Sticking coefficient β for silver deposited on gold and glass.

and beam fluxes of 10^{-4} to 10^{-2} monolayer/sec. It was found that β increased with increasing average deposit thickness on glass. The proposed explanation is the same as that outlined in the preceding paragraph. After one hour at a substrate temperature of 192°C and a beam flux of $\sim 10^{13}$ atoms/cm^2 sec, β assumed the values shown in Fig. 24. The decrease in β with increasing lattice misfit may have been due to an effect of σ_{c-x} according to equations (C-33) and (C-34). However, it is also possible that impurity adsorption, which would lower the nucleation rate by the model of Fig. 15, increased in the order gold, platinum, nickel.

Fraunfelder,[184] also working with a radioactive tracer and probably at low beam fluxes of 10^{-5} to 10^{-2} monolayer/sec, found $\beta = 0.3$ to 0.6 for deposition of silver on mechanically cleaned silver or mica substrates. However, β was 0.4 to 0.8 for condensation on a freshly evaporated silver substrate. In the case of cadmium the observed β was 0.0001 to 0.01 for deposition on mechanically cleaned glass, mica or polycrystalline metal substrates. On the other hand, for freshly evaporated silver, copper or gold substrates β

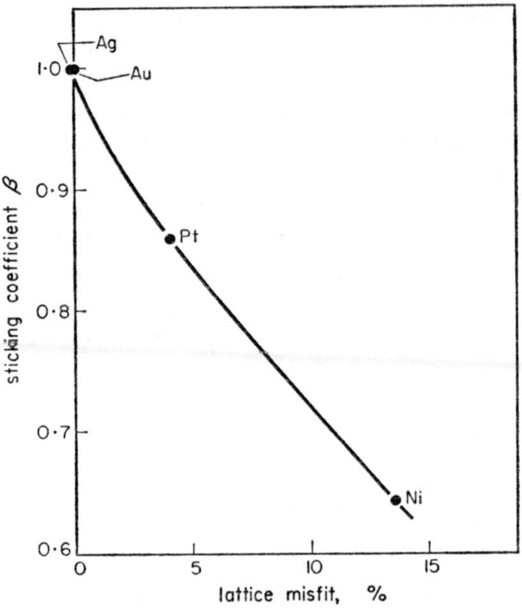

FIG. 24. Sticking coefficient β as a function of lattice mismatch for Ag at a flux of $\sim 10^{13}$ atoms/cm^2 sec incident on various polycrystalline substrates for one hour.

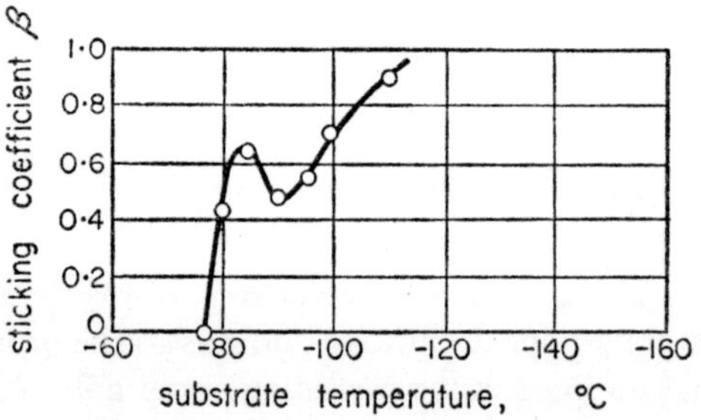

FIG. 25. Sticking coefficient at constant time (15 min) for mercury incident on nickel at substrate temperature T.

assumed values from 0.3 to 0.6. This work indicates that impurity adsorption has a major effect on the sticking probability.

Walther[185] measured the sticking coefficient of mercury on polycrystalline nickel at fluxes of $\sim 10^{16}$ atoms/cm^2 sec and found time-dependent β values which varied with substrate temperature as shown in Fig. 25. He postulated that: at the very low substrate temperatures the mercury atoms stuck wherever they impinged; at intermediate temperatures the diffusion coefficient of mercury on mercury facets was so small that some of the mercury re-evaporated before it could reach a growth site, accounting for the dip in the curve of Fig. 25*; at high temperatures the nucleation rate was small, giving a value of $\beta \simeq 0$. This experiment does not have quantitative significance but does illustrate the complexities that may be encountered in sticking coefficient determinations.

Other values of sticking coefficients have been determined, as reviewed by Wexler.[186] Also, Ehrlich[15] has reviewed work on sticking coefficients in the early stages of physical and chemical adsorption.

(b) *Other Pertinent Experimental Observations*

1. *Epitaxy*

Studies of epitaxial relationships between thin films or disperse deposits and substrates are pertinent to heterogeneous nucleation. The presence of epitaxy usually indicates that the nucleation rate is a maximum for nuclei of the epitaxial orientation.

There has been considerable confusion over the relation of nucleation theory to epitaxy. Accordingly, before considering the experimental results, the important predictions of section C-2 with respect to epitaxy are summarized as follows:

(a) Coherent nucleation is not required to produce epitaxy; epitaxy may also result from either semicoherent or non-coherent nucleation. The only requirement is that a particular orientation give lower interfacial free energies and hence a lower free energy of activation $\Delta G^\star$ and a much higher nucleation rate than any other orientation. For example, using the cap-nucleus model, $\phi_3(\theta)$ in equation (C-6) must be a minimum at the epitaxial orientation. It is true that coherency or semicoherency may contribute to the lowering of $\phi_3(\theta)$ through the mechanism discussed in connection with equations (C-33) and (C-34); however, such a contribution is not essential for epitaxy. In terms of the disc-nucleus model, the quantities ε_{c-v} and/or $(\sigma_{c-v} + \sigma_{c-x} - \sigma_{x-v})$ in equation (C-41) must be a minimum at the epitaxial orientation. Again, coherency or semicoherency may contribute to this minimization as considered in relation to equation (C-46).

(b) Coherent nucleation may occur only at high supersaturations where $|\Delta G_v| \gg C_3\delta'^2$ (equation C-34); otherwise semicoherent or non-coherent nucleation will occur. Thus, for a given beam flux, low substrate temperatures favor coherent nucleation.

(c) Even if coherent nucleation occurs, dislocations may enter the interface early in the growth process and make it semicoherent. This is directly analogous to the transition from coherency to semicoherency in the process of precipitate growth from solid solution.[187]

(d) At sufficiently high supersaturations many orientation relationships may give appreciable nucleation rates, thus destroying the epitaxy. Further,

* This part of his explanation seems highly unlikely.

at high supersaturations the many orientations generated in the rapid growth of polycrystals may obscure any epitaxial nucleation (see Section D).

(e) Finally, epitaxy may also result from crystal growth or from recrystallization of the deposit.

Thus one can predict with certainty that in cases where epitaxy is dependent only on nucleation there will always be a "transition" temperature, peculiar to the specific system of substrate and condensate, below which no favored orientation will be found. Further, this transition temperature will increase with beam flux. Finally, from the theory of Part C-2 it should be possible to make qualitative estimates of the relative values of these transition temperatures for various cases.

Seifert,[188] Bassett *et al.*[189] and Pashley[100] have recently reviewed epitaxy studies. The latter authors[189,190] conclude that coherent nucleation is unimportant in producing epitaxy and discount the interface models[147,148,191] proposed for coherent nucleation. Their conclusion is based on the observations that the crystallites are often isolated three-dimensional entities and that electron and X-ray diffraction studies fail to reveal coherency lattice strains in such crystallites. However, such observations are performed on crystallites which have grown a finite amount after nucleation. Thus in view of prediction (c) above, the evidence in regard to absence of strain does not suffice to exclude coherent nucleation. Also, a two-dimensional nucleus may grow into a three-dimensional crystallite. Hence occurrence of such crystallites does not exclude coherent nucleation.

Briefly summarizing the salient experimental results: A number of authors (e.g. references 171, 192–194) find evidence that high substrate temperatures are required for epitaxy of metal films and that below a critical temperature no favored orientation exists. Also, recent work[189,195,196] suggests that the transition temperature is not dependent only on condensate and substrate. In view of the predictions of nucleation theory it appears that supersaturation, substrate temperature and substrate cleanliness should also affect the transition temperature. Indeed, Sloope and Tiller[197] recently studied the deposition of 99.999 silver on (001) NaCl faces in a vacuum of 10^{-5} mm Hg. They found, in agreement with the above predictions, that the minimum substrate temperature for forming a single-crystal epitaxial film increased with increasing rate of deposition (i.e. supersaturation).

Shultz[198] studied the nucleation and growth of alkali halides on monocrystalline substrates of other alkali halides and produced electron micrographic evidence suggesting that two-dimensional nucleation occurs for mismatch $\delta' < 20\%$ but three-dimensional nucleation occurs for $\delta' > 20\%$. Bassett *et al.*[189] cite a few other cases of formation of two-dimensional crystallites. However, they note that in most cases the crystallites were three-dimensional. As pointed out above, the latter observations were made after some crystal growth had occurred and thus do not preclude two-dimensional nucleation.

Recently, Matthews[199] has observed by transmission electron microscopy the interface dislocations[147,148] that provide semicoherency at the interface between (001) planes of thin films of lead sulfide and lead selenide. These films are deposited from the vapor one on top of the other and are about 1000 Å thick. His analysis indicated that the dislocations had Burgers vectors of the type $\frac{1}{2}a\langle 110\rangle$ and that 96% of the lattice misfit was taken up by interface dislocations, i.e. in equation (C-33) $(\delta' - e)/\delta' = 0.96$. His observations were made some time after deposition took place so that nucleation may have occurred with a semicoherent interface or with a coherent interface which subsequently became semicoherent during or after growth. A similar observation of dislocations at the interface between some unknown precipitate and chromium bromide has been reported by Delavignette *et al.*[200]

Finally the work of Bryant *et al.*[201] is cited in which a connection between epitaxy and nucleation is clearly established. They placed aerosols of various crystalline substances in a cloud chamber to determine their catalytic potency in nucleating water and/or ice. They also studied heterogeneous nucleation from the vapor of ice on macroscopic crystals of these materials and obtained the same results for the relative catalytic potency. Further, it was found that the most potent catalysts all have epitaxial relations with ice which yield interfaces of small disregistry. Thus it appears that there may have been some degree of coherency in the nucleation.

From the foregoing discussion it appears that most of the qualitative aspects of epitaxy may be understood in terms of heterogeneous nucleation theory. However, in order to determine the exact quantitative relationships between epitaxy and heterogeneous nucleation it would seem necessary to study the epitaxy of deposits nucleated over a range of p, T, residual vacuum and substrate orientation. Also, in conjunction with a fit of observed $\Delta G_{v_{\mathrm{crit}}}$ to equations (C-20) or (C-45) it should be possible to distinguish between two- and three-dimensional nucleation.

2. *Amorphous Deposits and Metastable Crystals*

It was noted in Part C-2-(a)-6 that at very low temperatures adatom mobility is so low that an amorphous deposit may form. Whether such a deposit consists of tiny crystallites or not is immaterial. As long as there is no long-range order its properties will appear to be those of a liquid or amorphous substance. Semenov[202] suggested that heterogeneous nucleation always proceeds by formation of an amorphous film followed by nucleation of crystallites within the amorphous film. However, in view of the work of Moazed and Pound[145] it appears that such a possibility is likely for most metals only at very low temperatures* where $\Delta G^{\star}_{\mathrm{sd}} \gg kT$.

Buckel[203] found that copper, aluminum, lead, thallium, mercury, zinc and tin all form crystalline thin films in deposition from the vapor on collodion

* By "low" temperature we mean low with respect to a characteristic temperature for atomic mobility. Hence it is a material property.

substrates at temperatures down to 2°K. Bismuth forms amorphous films at 2°K $< T <$ 15°K and crystalline films above 15°K; gallium behaves similarly. Germanium and silicon form amorphous films to approximately room temperature. He also observed that at 20 to 80°K a metastable high-temperature phase of gallium deposits and that metastable white tin is formed at low temperatures, in agreement with Ostwald's "law of stages"[(152)] [Part C-2-(a)-(7)].

Götzberger[(204)] observed that antimony which has been deposited on glass at −50°C to room temperature nucleates and grows as spherical segments. These are amorphous when the average deposit thickness is < 200 Å but crystalline (hexagonal) when this thickness is > 200 Å. Palatnik and Komnik[(205)] studied the deposition of bismuth on a glass substrate at 30°C $< T <$ 250°C. They found that, at a given supersaturation, crystalline bismuth nucleates below a critical temperature. Above this temperature liquid nucleates and grows and later transforms to solid. Again, this is indicative of the effect of capillarity on metastable-phase nucleation as discussed in Part C-2-(a)-(7) in connection with Ostwald's "law of stages". This effect is readily explained in terms of elementary nucleation theory: at a sufficiently low temperature the increase in $\Delta G_{v_{crit}}$ obtained by nucleation of crystals instead of liquid outweighs the increase in σ_{c-v} and/or $\phi_3(\theta)$ (e.g. equation C-6); this results in a lower $\Delta G^{\star}$ and a faster nucleation process. Also, they plotted the critical pressures for nucleation of liquid and solid as a function of $1/T$ and noted a similarity to the ordinary bismuth phase diagram as shown in Fig. 26.

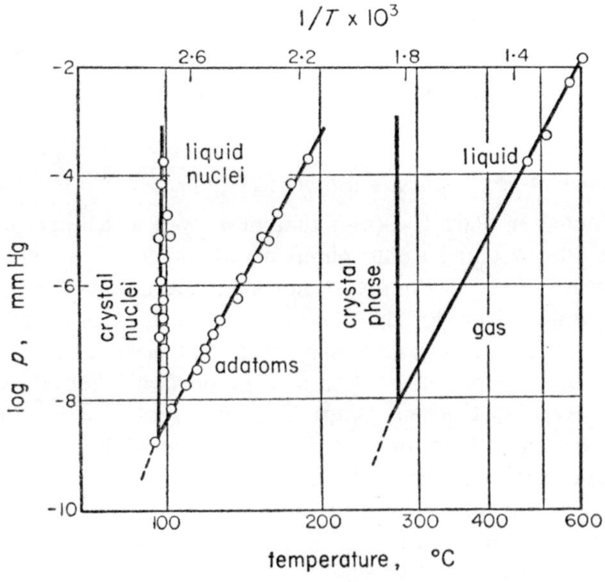

FIG. 26. Critical pressure for nucleation of bismuth on a glass substrate and bismuth phase diagram.

In a similar investigation Bryant *et al.*[201] studied the epitaxial deposition of ice on substrates and found that below −6°C ice nucleated and grew as platelets at macroscopic ledges on the basal plane on PbI_2. Below −20°C platelets formed at macroscopic ledges on basal planes of CdI_2. Similar results were obtained for AgI, CuS, I_2, HgI_2, calcite, muscovite and brucite. In two cases, AgI and CdI_2, the appearance of ice was studied as a function of supersaturation and substrate temperature. The results given in Fig. 27 for

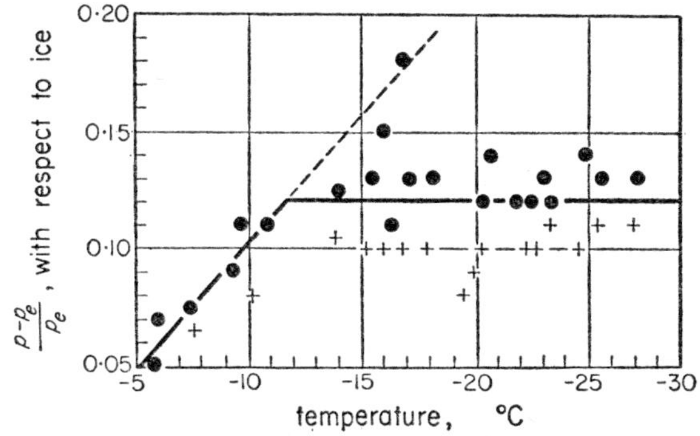

● ice crystal formation + no precipitation
45° line corresponds to saturation with respect to water

FIG. 27. Conditions of temperature and supersaturation at which oriented ice crystals appeared on AgI crystal.

AgI indicate that at temperatures below −12°C ice crystals appeared when the air was undersaturated with respect to water but supersaturated with respect to ice by 12%. On the other hand at temperatures between −4 and −12°C ice appeared only if the air was supersaturated with respect to water. Thus between −4 and −12°C the initial deposit may have been liquid water, which froze to ice during subsequent growth.

3. *Nucleation on Lattice Imperfections*

In the work of Bryant *et al.*[201] it was noted that the ice nucleated at step edges at supersaturations of 10 to 20% while nucleation of ice occurred on the flat low-index surfaces only at much higher supersaturations (~100%). Andrade and Martindale[206] formed thin films of silver and gold by sputtering and produced micrographs which indicate preferential nucleation at cracks in quartz or glass substrates. Metz and Lad[207] deposited gold, silver and zinc on cleaved low-index substrates of NaCl or KBr. They observed preferential nucleation at scratches and ahead of and along macroscopic cleavage cracks. Similar observations were made by Fray and Nielsen[176] for zinc on glass. The preferential nucleation at scratches and cracks may be explained in terms of the theory of Part C-2-(a)-(4).

If the cleaved alkali halide crystals[207] were annealed at $0°C < T < 200°C$ prior to deposition, the nuclei formed in rows. It was also noted that the number of nuclei was greater when the substrate has been irradiated. In view of these facts and Gilman's work on dislocations in LiF[208] it seems likely that the nucleation on the low-index planes away from the cracks and scratches was catalyzed by dislocations.

Mossop,[209] using aerosols of $CaCO_3$ and $CaSO_4 \cdot 2H_2O$, and D'Albe,[210] using aerosols of CdI_2, investigated the potency of catalysts for condensation of water and ice in cloud chambers. They observed that only water droplets nucleated at terminal temperatures of -4 to $-24°C$. However, they found that if the catalyst particles were pre-cooled by large expansions to terminal temperatures of -24 to $-40°C$ where ice appeared in the cloud chamber and then were allowed to warm to somewhat above 0°C where the macroscopic ice melted, subsequent expansions produced ice nuclei at -4 to $-24°C$. This indicates that ice was retained in microcavities in the catalyst particles and subsequently acted as a potent catalyst for ice nucleation as discussed in Part C-2-(a)-(4).[150]

Bassett[189,211] examined NaCl cleavage surfaces on which gold had been deposited and found that preferential nucleation of gold crystallites occurs at monomolecular steps. On the other hand, this decoration of the steps by gold particles was not observed in deposition of gold on mica cleavage surfaces. It is believed that these phenomena may be understood in terms of the theory of Chakraverty and Pound[151] [Part C-2-(a)-(4)].

4. *Effect of Thermal Velocity of Incident Atoms*

As discussed in Parts A and C-3-(a)-(2), incident atoms should undergo rapid thermal equilibration with the substrate, i.e. within the period of a few vibrations on the surface. Further, they should stick to the surface for more than a few vibrations in cases where $\Delta G^{\star}_{des} \gtrsim kT$. Sears and Cahn[47] suggested that in the experiment of Yang *et al.*[168] adatoms may have retained some of their incident kinetic energy and hence re-evaporated at a rate greater than would be expected for the substrate temperature. However, considering the large values of $\Delta G^{\star}_{des}$ (Tables 5 and 6) calculated from the experimental data, the reasoning of Part A predicts a negligible loss of adatoms by such a mechanism.

Aziz and Scott[212] deposited silver on glass from a high-energy direct beam and from a low-energy beam which had been reflected from teflon or diffused through nitrogen. They found a greater number of nuclei for a given amount of deposit in the case of the low-energy beam. The results were interpreted as showing that the adatoms retain some kinetic energy on adsorption. Holland[213] pointed out that the substrate temperature probably varied in the above test and that the presence of the residual gas could affect nucleation frequency. Although Aziz and Scott[214] feel that the effects suggested by

Holland are small, their results do not unequivocally demonstrate retention of kinetic energy by adatoms.

Thus while effects due to retention of kinetic energy have been suggested, none have been clearly demonstrated.

5. *Effect of Beam Orientation*

Evans and Wilman[215] for iron on glass and Melnikova *et al.*[216] for magnesium, cadmium and zinc on glass and copper* found that as a film grown from an atomic beam thickens it develops a preferred orientation which is related to the angle of incidence of the beam. This may be a crystal-growth effect. At any rate the result is not readily understood from nucleation theory.

6. *Loss of Adatoms by Diffusion into the Substrate*

Chirigos *et al.*[160] for silver on polycrystalline copper at 400°C, Glossop and Pashley (unpublished, see ref. 189, p. 14) for copper on polycrystalline gold at room temperature and Siegel and Peterson[217] for copper on carbon at 350°C have demonstrated that, under the cited conditions, adatoms were lost at an appreciable rate by diffusion into the substrate.

4. Summary of Experiment and Conclusions

Almost all of the many semi-quantitative observations (e.g. sticking coefficient measurements) relating to nucleation from the vapor of crystals on substrates are consistent with the general theory which assumes metastable equilibrium between single adatoms and polyatomic embryos on the substrate surface and ascribes macroscopic thermodynamic properties to these small clusters. The evidence is overwhelming that the adatoms are mobile on the substrate and that they may re-evaporate unless the temperature is low.

Unlike the situation in the case of homogeneous nucleation in the vapor, large statistical mechanical contributions to the free energy of formation which arise from translation and rotation of the embryos are likely only for nucleation at high temperatures on substrates to which their binding energy is small. The appreciable statistical contribution given by equation (C-14) or (C-21) is similar to that of translation in homogeneous nucleation from the vapor and appears to have always been neglected in the past. Also unlike the case of nucleation in the vapor, it appears that there are no large problems relating to thermal non-accommodation or transients in nucleation on substrates.

Only two investigations,[167,168] in which critical supersaturations were measured as a function of substrate temperature, have provided a quantitative test to prove the usefulness of the recommended theoretical equations

* Polycrystalline.

(C-12), (C-20) and (C-26). Even in these cases the exact role of impurity adsorption was not determined. It is noteworthy that only one quantitative nucleation experiment[145] has been conducted on a perfectly clean monocrystalline metallic substrate. However, this work has not yet been extended to cover a range of substrate temperatures.

In general the supersaturation required for appreciable nucleation rate is quite high and strongly dependent on both condensate and substrate. In the usual cases of non-metallic substrates or contaminated metallic substrates, the principal impediment to nucleation of metallic crystals seems to be the interfacial free energy requirement for formation of the critical nucleus. The effect of adsorbed contaminants usually is to reduce the catalytic potency of the substrate for nucleation.

It appears that nucleation occurs preferentially on non-close-packed planes of the substrate. Evidently, lattice disregistry between condensate and substrate is not an important factor in increasing the thermal activation barrier to nucleation.

In view of the scarcity of significant quantitative data in this field it would seem that almost any critical supersaturation measurement in which crystallographic orientation, degree of imperfection and extent of contamination of the substrate are known and controlled should be most fruitful.

The authors feel that most of the salient qualitative results of studies on epitaxy may be understood (indeed, could have been predicted) from the nucleation theory outlined in Part C-2. Similarly, it would appear that this theory affords a reasonable description of the deposition of amorphous films and metastable crystals.

On the other hand, much work remains to be done on the kinetics of nucleation at imperfections on substrate surfaces. Also, the moot effect of thermal velocity of incident atoms should receive further experimental attention.

D. GROWTH AND EVAPORATION OF LIQUIDS AND DISLOCATION-FREE CRYSTALS

1. Introduction

THE ideas developed in the foregoing are applied to crystal growth in this section. Simple liquids should exhibit no interface control in evaporation or growth. As a result the kinetics of evaporation or growth are simplest for such liquids and are considered first and shown to follow the ideal equation (A-2). In sequence, deviations from ideality are treated which may arise because of entropy constraints, heat-flow constraints and/or diffusion constraints.

Dislocation-free crystals are then considered. A distinction is made between singular and non-singular solids. Non-singular solids follow the same kinetic laws as liquids. The growth of singular solids is treated on the basis of nucleation theory as developed in Section C. Distinctions between the kinetics of growth and evaporation are developed. Finally, microscopic kinetics and macroscopic surface topography changes are related. In each case current theories and experiments are cited.

2. Growth and Evaporation of Liquids

(a) *Ideal Growth*

1. *Theory*

Simply bonded liquids which evaporate or grow by exchange of single atoms with a monatomic vapor should exhibit no surface constraint in the kinetics because each surface site may be considered equivalent and because, as noted in Section A, elastic reflection of incident atoms is unlikely.* For such a liquid, every atom arriving at the surface should condense and the ideal growth rate equation (A-4) with $\alpha_c = \alpha_{c_1} = 1$ should obtain. Except at very high beam fluxes, of the order of 10^{25} atoms/cm^2/sec, there is little direct interference of condensation and evaporation fluxes, i.e. few impinging atoms collide with surface atoms which are activated for evaporation. In

* $\alpha_{c_1} = 1$.

this usual situation, the gross evaporation flux is the same as that under equilibrium conditions and the evaporation coefficient $\alpha_{v_2} = \alpha_{c_2} = 1$.*†

Even though the above argument indicates $\alpha_{v_2} = 1$, there has been considerable interest in developing a mechanistic expression for the gross evaporation flux. This problem was introduced by Polanyi and Wigner[218] and Herzfeld[219] and has since been considered by a number of authors.[220–224] All of these treatments in essence use the approach of absolute rate theory and describe the evaporation frequency by the product of a frequency of decomposition $\nu^\star$ and the probability of occupancy of the activated state. Such an approach yields[6] the gross flux of evaporation‡

$$J_{v_g} = \nu^\star n^\star = n_L \nu^\star \exp(-\Delta G^\star_{\text{vap}}/kT) = n_L \nu^\star \frac{F^\star}{F} \exp(-\Delta E_{0_{\text{vap}}}/kT) \quad \text{(D-1)}$$

where $n^\star$ and n_L are the concentrations of atoms in the activated state and reactant state, respectively, $F^\star$ and F the partition functions for the activated state and reactant state, respectively, $\Delta G^\star_{\text{vap}}$ is the Gibbs free energy of activation, and $\Delta E_{0_{\text{vap}}}$ the potential energy difference between reactant and activated states, in this case approximately equal to the energy of evaporation.§ The above mentioned authors,[218–224] however, all differ in describing the activated state and hence in assigning values to $F^\star$, $\nu^\star$, etc. As noted in Section A, we feel that the current state of knowledge of electronic and ionic interaction of an atom with a surface justifies only the simplest possible approach to the description of the activated state. Eyring *et al.*[6] factor a partition function $f^\star_{\text{tr}}$, corresponding to a translational|| degree of freedom in the direction of evaporation, from $F^\star$ in equation (D-1). Wert and Zener[140] retain an equivalent number of degrees of freedom in the descriptions of reactant and activated states by factoring another partition function f_V, relating to a vibrational degree of freedom in the direction of activation, from F. The remaining transverse vibrational and/or translational degrees of freedom are regarded as unchanged in the reactant or activated states. Also it is assumed that $f^\star_{\text{tr}}/f_V \simeq 1$ so that $\Delta G^{\star\prime\prime}_{\text{vap}}$, the Gibbs free energy difference between reactant and activated states as described with a degree of freedom

* The major interest in this book is in the evaporation and growth of solids; liquids are considered in order to introduce the various constraints for a simple system, and then the more complicated kinetics for solids are considered. Hence we here deal only with the evaporation of liquids under conditions where normal evaporation occurs from a free surface. Evaporation under conditions of ebullition (boiling) is treated, e.g. by Carman.[225]

† The subscript 2 refers to the coefficient occasioned by direct interference.

‡ Note that if one were considering the final stage of evaporation to consist of desorption from an adsorbed layer, the same equation would apply except that the reactant state would be the adsorbed state and $\Delta G^\star_{\text{vap}}$ would be replaced by $\Delta G^\star_{\text{des}}$.

§ Actually the energy of evaporation $\Delta E = -Nk[\text{d} \ln (F_{\text{vap}}/F)/\text{d}(1/T)] + \Delta E_{0_{\text{vap}}}$ where F_{vap} is the partition function in the vapor; however, the temperature dependence of F_{vap} and F are small so one uses the approximation $\Delta E \simeq \Delta E_{0_{\text{vap}}}$.

|| A more approximate treatment which gives an equivalent result but which avoids consideration of the "width" of the activation barrier is given by Laidler.[226]

missing from each, equals the normal thermodynamic free energy of activation $\Delta G^{\star}_{\text{vap}}$. Thus from equation (D-1)

$$J_{v_g} = n_L \nu^{\star} \frac{f^{\star}_{\text{tr}}}{f_V} \exp(-\Delta G^{\star''}_{\text{vap}}/kT) \tag{D-2}$$

$$\simeq n_L \nu^{\star}[(2\pi mkT)^{\frac{1}{2}}\bar{\delta}/h]/[1 - \exp(-h\nu/kT)]^{-1} \exp(-\Delta G^{\star}_{\text{vap}}/kT) \tag{D-3}$$

where $\bar{\delta}$ is the "width" of the top of the activation barrier in the direction of decomposition and ν the Einstein vibrational frequency in the reactant or normal state. Now[6] $\nu^{\star} = (\dot{x}/\bar{\delta}) = (kT/2\pi m)^{\frac{1}{2}}/\bar{\delta}$, where $\dot{x}$ is the mean velocity of the activated species in the direction of decomposition, so that equation (D-3) becomes

$$J_{v_g} \equiv n_L \bar{\omega} \exp(-\Delta G^{\star}_{\text{vap}}/kT) \tag{D-4}$$

in which

$$\bar{\omega} \equiv (kT/h)[1 - \exp(-h\nu/kT)] = \nu^{\star}. \tag{D-5}$$

In many cases $\nu < (kT/h)$ and the exponential containing ν may be expanded; in this situation $\bar{\omega}$ is identical with ν as well as $\nu^{\star}$. Hence the gross flux of evaporation under equilibrium conditions may be given by

$$J_{v_g} = n_L \nu \exp(-\Delta G^{\star}_{\text{vap}}/kT) \equiv p_e/(2\pi mkT)^{\frac{1}{2}} \tag{D-6}$$

where n_L is the surface density of atoms in the liquid. Under non-equilibrium conditions, as noted above, the gross evaporation flux will be unchanged, while the gross condensation flux will equal $p/(2\pi mkT)^{\frac{1}{2}}$. Thus the net flux under non-equilibrium conditions will be that given by the ideal equation (A-2).

Liquids of spherically symmetric molecules which do not have appreciable entropies of activation should follow kinetics similar to those of atomic liquids and should also exhibit an $\alpha_{c_2} = \alpha_{v_2} = 1$.

2. *Experiment*

Most experiments determine α_{v_2} by comparing free evaporation rates with the rate expected from the ideal equation (A-2). Knudsen,[227] in the first careful measurement of this quantity, determined that $\alpha_{v_2} = 1$ for clean liquid mercury. He found it necessary to continuously expose a fresh clean surface of the mercury in order to avoid contamination and the attendant apparent lowering of α_{v_2}. Neumann and Schmoll[228] measured the free evaporation of liquid potassium and by comparing their results with the equilibrium pressure[229] found that $\alpha_{v_2} = 1$. Volmer and Estermann[230] confirmed Knudsen's result in finding $\alpha_{v_2} = 1$ for liquid mercury. Holden *et al.*[231] find $\alpha_{v_2} = 1$ for liquid beryllium. Also for CCl_4, which is a spherically symmetrical molecule, Alty and Mackay[51] and others[232,233] find $\alpha_{v_2} = 1$ for the liquid. One notes that the surface temperature was accurately known in each of these experiments.

In conclusion, for clean surfaces of liquids which are undergoing normal evaporation, α_{v_2} equals unity in agreement with theoretical prediction. By implication, α_{c_2} equals unity also.

(b) *Entropy Constraints*

1. *Theory*

Consider again equation (D-2). Suppose, following Eyring and others,[219–224] that a molecule is restricted in a rotational degree of freedom, say as a torsional vibration, in the liquid state but not in the vapor. Then one may write

$$J_{v_g} = n_L \nu^\star \frac{f_{\text{tr}}^\star}{f_v} \frac{f_R^\star}{f_{R_L}} \exp(\Delta G_{\text{vap}}^{\star\prime}/kT) \tag{D-7}$$

where $f_R^\star$ and f_{R_L} are the rotational partition functions in the activated and restricted liquid states, respectively, and $\Delta G_{\text{vap}}^{\star\prime}$ is the activational free energy with the rotational and some of the other degrees of freedom removed as before. The temperature dependence of the terms $f_R^\star$, f_{R_L}, etc., are small. Accordingly one considers the f_R terms to affect mainly $\Delta S_{\text{vap}}^\star$. Hence, neglecting their effect on $\Delta H_{\text{vap}}^\star$, supposing that contributions to $\Delta S_{\text{vap}}^\star$ are small for the terms other than the rotational terms, and making the same expansions and approximations that lead to the final form of equation (D-4), one obtains

$$J_{v_g} = n_L \bar{\omega}(f_R^\star/f_{R_L}) \exp(-\Delta H_{\text{vap}}^\star/kT). \tag{D-8}$$

In regard to condensation of, say, a polar molecule, we refer back to the model of Fig. 1. The activated state for condensation into the liquid may be considered a vapor molecule just impinging on the surface. If improperly oriented the molecule will be repelled, while if properly oriented the molecule will be attracted to the surface and will condense. Hence the gross condensation flux should be approximately

$$J_{c_g} = (f_R^\star/f_{R_v})p/(2\pi mkT)^{\frac{1}{2}} \tag{D-9}$$

where f_{R_v} is the rotational partition function in the vapor phase. As in the case of simple liquids, there is little likelihood of direct interference of condensation and evaporation fluxes. Thus J_{v_g} under non-equilibrium conditions should equal J_{v_g} under equilibrium conditions. Further, $J_{v_g} = J_{c_g}$ at equilibrium. Accordingly, comparing equations (D-8) and (D-9), the net flux of condensation or evaporation is given by

$$J_i = J_{c_g} - J_{v_g} = (f_R^\star/f_{R_v})(p - p_e)/(2\pi mkT)^{\frac{1}{2}} \tag{D-10}$$

leading to

$$\alpha_{v_3} = \alpha_{c_3} = (f_R^\star/f_{R_v}). \tag{D-11}$$

Referring to equations (D-8) and (D-9) it is seen that two possibilities exist: either (a) the rotational degrees of freedom are operative in the activated state, $f_R^\star = f_{R_v}$, and

$$\alpha_{v_3} = \alpha_{c_3} = 1 \tag{D-12}$$

or (b) the rotational degrees of freedom are restricted to torsional vibrations in the activated state, $f_R^\star \simeq f_{R_L}$, and

$$\alpha_{v_3} = \alpha_{c_3} = \delta < 1, \tag{D-13}$$

where δ is the "free angle" ratio introduced by Stearn and Eyring[220] and is equal to f_{R_L}/f_{R_v}. The latter possibility seems the more likely.

Wyllie[234] has suggested that other degrees of freedom, such as vibrational degrees of freedom for the molecule in the vapor phase, may also enter δ. However, Mortensen and Eyring[224] note that the relaxation time to activate a rotational degree of freedom in the activated state should be longest and hence δ should be most strongly affected by rotational terms. Since the latter view favors the simple interpretation of δ given above, δ shall be considered to contain only the terms f_{R_v} and $f_R^\star$. A more complicated interpretation of δ may be justified when experimental tests become more exact.

Wyllie[234] has considered the possibility that when surface polarization is important evaporation and condensation may proceed via a two-dimensional gas or adsorbed layer.* In such a case the kinetics would involve

$$\text{liquid} \underset{J_{L_2}}{\overset{J_{L_1}}{\rightleftharpoons}} \text{adsorbed} \tag{D-14}$$

$$\text{adsorbed} \underset{J_{L_4}}{\overset{J_{L_3}}{\rightleftharpoons}} \text{vapor} \tag{D-15}$$

In this two-stage process, the entropy constraints are supposed to enter the liquid-adsorbed state kinetics. Writing down the kinetic equations for each step in equations (D-14) and (D-15),

$$J_{L_1} = n_L \bar{\omega} (f_{R\text{ad}}^\star / f_{R_L}) \exp(-\Delta H_{L_1}^\star / kT), \tag{D-16}$$

$$J_{L_2} = n_L a^2 n_s \bar{\omega} (f_{R\text{ad}}^\star / f_{R\text{ad}}) \exp[-(\Delta H_{L_2}^\star + \Delta H_{\text{hole}})/kT], \tag{D-17}$$

$$J_{L_3} = n_s \bar{\omega} (f_R^\star / f_{R\text{ad}}) \exp(-\Delta H_{L_3}^\star / kT), \tag{D-18}$$

and

$$J_{L_4} = (f_R^\star / f_{R_v}) p / (2\pi m k T)^{\frac{1}{2}}, \tag{D-19}$$

where n_L and n_s are the concentrations of atoms in the liquid surface and the adsorbed state, respectively, a is the diffusion jump distance on the surface,

* Whether one regards a molecule as translating in a two-dimensional gas or diffusing as an adsorbed molecule depends on the magnitude of the activational free energy for surface diffusion $\Delta G_{\text{sd}}^\star$ compared with kT, as noted in the discussion preceding equation (C-9).

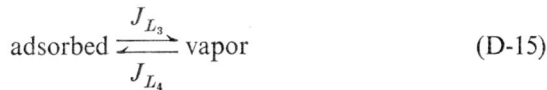

$f^{\star}_{R_{ad}}$ and $f_{R_{ad}}$ are the rotational partition functions in the activated state between the liquid and the adsorbed layer and in the adsorbed layer, respectively, the $\Delta H^{\star}$ terms are enthalpies of activation and the ΔH_{hole} term reflects the probability that a "hole" in the liquid surface is available for adsorbed atoms to drop into. It is likely that metastable equilibrium of the adsorbed state will exist so that n_s will be a constant. Noting that $J_{L_4} - J_{L_3} = J_{L_2} - J_{L_1}$ and $a^2 \simeq 1/n_L$, one obtains

$$\alpha_{v_4} = \alpha_{c_4} = (f^{\star}_R/f_{R_v})/\{1 + (f^{\star}_R/f^{\star}_{R_{ad}}) \exp [(\Delta H^{\star}_{L_2} + \Delta H_{\text{hole}} - \Delta H^{\star}_{L_3})/kT]\} \quad \text{(D-20)}$$

which is somewhat similar to the result for the one-step process treated above.

2. *Experiment*

Experiments pertinent to entropy constraints are summarized in Table 8. It can be seen that the agreement of the results in Part I of the table with equation (D-13) is good in all cases. However, except for the case of benzene, agreement with equation (D-20) is possible. Neumann,[221,228] Penner,[222] and most recently Mortensen and Eyring[224] concluded that the agreement in Part I of the table confirms equation (D-13). Considering the care taken in the experiments to avoid contamination or errors in surface temperature, we concur with the above authors in interpreting the results as due to an entropy constraint.

Wyllie,[234] though he derived a relationship of the form of equation (D-20), concluded that an energy of activation for condensation accounted for the low α_{v_3} values in Part I of the Table 8. Such an energy term might arise from a change in bond length in going from the vapor to the activated state for evaporation or condensation. Littlewood and Rideal[239] concluded that the above results are due to errors in measuring surface temperatures occasioned by large temperature gradients at the surface. Either of these latter interpretations would lead to $\alpha'_T \neq 1$* (see Part A). The principal evidence, then, favoring the δ interpretation of the low α_{v_3} values, other than the agreement between α_{v_3} and δ, is in the finding of Alty and Mackay[51] that $\alpha'_T = 1$ in evaporation of water even though $\alpha_{v_3} = 0.04$. According to the theory outlined above, $\alpha_{c_3} = \alpha_{v_3} = \delta$ in these cases.

The remaining results in the table do not follow equation (D-13). Hickman[237] and Trevoy[236] measured the evaporation coefficient of water and glycerol, respectively, as the liquid flowed in a stream emerging at high velocity from an orifice and found α_v to be near unity. On the other hand Hickman and Torpey[243] found for still water that $\alpha_v = 0.001$ to 0.02, depending on the purity of the water. They concluded that clean water really has an $\alpha_v = 1$ and that the results in Part I of Table 8 are all due to impurity adsorption at the evaporating interface. However, from Hickman's[237] results we calculate that the water stream in his experiment was moving

* $\alpha'_T = T_v/T_L$.

TABLE 8

α_{v3} AND δ FOR VARIOUS POLAR COMPOUNDS

Substance	α_{v3}	δ	Ref.
I. Benzene	0.9	0.85	233
Chloroform	0.16	0.54	233
Ethanol	0.020	0.02	233
	0.024	0.02	235
Methanol	0.045	0.05	233
Glycerol	0.05	0.05	234
Water	0.04	0.04	
	0.02	0.04	232
II. Glycerol	1	0.05	236
Water	0.4	0.04	237
III. 2-ethyl-hexyl phthalate	1	—	238
2-ethyl-hexyl-sebacate	1	—	238
Myristic acid	1	—	239
Lauric acid	1	—	239
Hexadecanol	1	—	239
Capric acid	0.49	—	239
Tetradecanol	0.68	—	239
Dodecanol	0.22	—	239
n-di-butyl phthalate	1	—	240
Tri-heptymethane	1	—	241
Tridecyl methane	1	—	241
n-hexadecane	1	—	242
n-octadecane	1	—	242
n-heptadecane	1	—	242

at $\sim 10^3$ cm/sec. Hence we feel that turbulence of the stream could continuously disturb the liquid–vapor interface and thus disrupt the surface dipole.(244) The tentative conclusion is that the results in Part II of Table 8 are valid for a disturbed and hence unpolarized liquid–vapor interface; for an undisturbed, polarized liquid–vapor interface, the results of Part I of the table obtain. Further experiments to clarify this point seem desirable.

Finally, there are the results listed in Part III of Table 8. In several cases(239–242) corrections were made of surface temperatures to show that $\alpha_v = 1$ although the apparent value of α_v was less than one. Even in the cases of capric acid, tetradecanol and dodecanol, Littlewood and Rideal(239) feel that the true value of α_v is unity and that deviations were due to surface cooling. Hence although the substances listed in Part III of the table are polar, it appears that $\alpha_v = 1$. Mortensen and Eyring(224) note that all of these substances are either long-chain molecules or large planar molecules and suggest that rotational degrees of freedom are operative in the activated state, a circumstance which by the above arguments would lead to equation (D-12) or $\alpha_{v_3} = 1$.

In conclusion, equation (D-13) (or equation D-20) holds for small polar molecules evaporating from static surfaces; equation (D-12) holds for polar molecules which are of the long-chain type or are large and planar. Some evidence has been interpreted as indicating that either surface contamination or surface cooling actually is the cause of the low α_{v_3} values in all cases. In one case this latter interpretation was shown to be tenuous by concurrently measuring α_{v_3} and α'_T. Similar experiments, in which α_{v_3} and α'_T or α_{c_3} and α'_T are measured concurrently, should clarify the role of δ in evaporation and growth.

(c) *Surface Contamination*

1. *Theory*

For clean liquid metals α_v should equal unity as noted above. Contamination of such surfaces should lower the condensation or evaporation probability by imposing an interface constraint. In the limit, diffusion through an adsorbed film could be rate controlling, and the rate of growth or evaporation could be much less than the ideal rate given by equation (A-2). The detailed kinetics are not treated here but several possibilities exist including: reflection of incident atoms because of a lowering of $\Delta G^{\star}_{\text{des}}$, slow diffusion through impurity adsorbate, and slow dissociation of an atom from the liquid at the adsorbate interface, all of which would lead to

$$\alpha_{v_5} = \alpha_{c_5} < 1. \tag{D-21}$$

For more complex molecules where $\alpha_v < 1$ due to δ effects, it is conceivable that adsorption of an impurity might increase α_v by reducing the δ constraint. However, it seems more likely, even in the case of such complex molecules, that the principal effect of adsorption will be to lower α_{v_5} and α_{c_5} as discussed in the preceding paragraph.

2. *Experiment*

Knudsen[227] found that contamination of a liquid mercury surface by impurities adsorbed from a poor vacuum reduced the vaporization coefficient from unity to an α_{v_5} as low as 10^{-3} (see also reference 245). In general, once the role of impurity adsorption on α_{v_5} and α_{c_5} was appreciated, attempts were made to avoid contamination in measurements of evaporation or growth rates.

(d) *Surface-Temperature Effects*

As noted by Bradley and Shellard[242] and Littlewood and Rideal,[239] "self-cooling or heating" may result in a surface temperature different from the measured temperature of the bulk liquid. Such a circumstance is favored by high net fluxes and/or low thermal conductivity of the liquid. These authors considered experiments where surface-temperature corrections are necessary in the evaluation of evaporation coefficients. Neglect of the surface-temperature correction leads to low apparent values of either condensation or evaporation coefficients.

(e) *Diffusion in the Vapor Phase*

1. *Theory*

The evaporation or growth of a liquid in the presence of an inert gas has been considered by Stefan,[246] Mache,[247] Langmuir,[248] Fuchs,[249] Bradley *et al.*,[250] Knacke and Stranski,[251] Monchick and Reiss,[252] Frisch and Collins,[253] and Kirkaldy.[254] The net rate of evaporation or growth is given by

$$J_v = -D_v \nabla n_v \tag{D-22}$$

where n_v is the concentration of the evaporating substance in the vapor phase. It is supposed[249,250] that:* a steady-state concentration $n_v(\Delta)$ is maintained at a distance Δ, about equal to the mean free path in the vapor phase, from the liquid interface; steady-state diffusion occurs between Δ and L, the position of a sink for the vapor where a concentration $n_v(L)$ is maintained; normal evaporation and condensation kinetics obtain at the liquid interface and any α_{v_i} terms at the liquid interface are unaffected by the diffusion constraint.† Thus for unidirectional diffusion, equation (D-22) becomes

$$J_v = -\alpha_{v_i}(kT/2\pi m)^{\frac{1}{2}}[n_{v_e} - n_v(\Delta)] = -\frac{D}{(L-\Delta)}[n_v(L) - n_v(\Delta)], \tag{D-23}$$

or, eliminating $n_v(\Delta)$,

$$\begin{aligned} J_v &= -\frac{\alpha_{v_i}(kT/2\pi m)^{\frac{1}{2}}[n_{v_e} - n_v(L)]}{1 + \alpha_{v_i}(kT/2\pi m)^{\frac{1}{2}}(L-\Delta)/D} \\ &= \frac{\alpha_{v_i}[p(L) - p_e]/(2\pi mkT)^{\frac{1}{2}}}{1 + \alpha_{v_i}(kT/2\pi m)^{\frac{1}{2}}(L-\Delta)/D}. \end{aligned} \tag{D-24}$$

The equivalent expression for the evaporation or growth of a spherical droplet of radius r by diffusion through a concentric spherical shell of thickness $(R - \Delta)$ has been shown[250] to be

$$J_v = \frac{\alpha_{v_i}[p(R) - p_e]/(2\pi mkT)^{\frac{1}{2}}}{1 + \alpha_{v_i}(kT/2\pi m)^{\frac{1}{2}} r^2 (R - r - \Delta)/DR(r+\Delta)}. \tag{D-25}$$

In the limit of high pressure of residual inert gas (i.e. small D and Δ), equations (D-24) and (D-25) reduce to the simpler Langmuir[248] diffusion equation

$$J_v = -[D/(L-\Delta)][n_{v_e} - n_v(L)], \tag{D-26}$$

or

$$J_v = -[DR/r(R-r)][n_{v_e} - n_v(R)]. \tag{D-27}$$

At the limit of low inert gas pressure the denominators of equations (D-24) and (D-25) become unity, and the simple free-evaporation formulae obtain.

* Monchick and Reiss[252] and Frisch and Collins[253] have shown that these assumptions describe a limiting case which can be developed from fundamental statistical mechanical considerations.

† We note a reservation in regard to the last assumption. If α_{v_i} contains δ effects, the apparent value of $\alpha_{v_i} = \delta$ could be greater than the true value measured in free evaporation. For example, the presence of the inert gas might affect the local equilibrium in the activated state.

Evidently an apparently low value of α_{v_i} will be obtained if important diffusion constraints are neglected in analyzing an experiment. Finally, it is noted that the "inert" gas may adsorb at the liquid–vapor interface and slow the kinetics through an α_{v_5}, α_{c_5} term.

2. *Experiment*

Of course the transpiration method, which involves diffusive flow in the vapor phase, is used extensively in measurements of equilibrium vapor pressures.[255,256] We shall not discuss such experiments and merely note that the effect of diffusion is compensated for in these cases.*

Knacke and Stranski[251] cite the work of Mache[247] and of Gudris and Kulikowa[257] who measured the evaporation rates of water in the presence of hydrogen, hydrogen–air mixtures, nitrogen and carbon dioxide. They found that the rates of evaporation were slower in the presence of the other gases than for hydrogen, as would be expected in view of the Maxwell expression for the diffusion coefficient in a binary gaseous system.[1]

A clear application of the formulae (D-24) to (D-27) is given in the work of Bradley *et al.*[240,250] who measured the evaporation rates of small droplets of di-butyl phthalate in various pressures of air, hydrogen or freon at 20°C. They plotted J_v versus the reciprocal of p_{inert}, the pressure of the inert gas,† and fitted the high-pressure extrapolation of the curve to equation (D-27) to obtain a value for D. Then by calculating Δ from gas-kinetic theory, they succeeded in obtaining α_v from the experimental data and equation (D-25). They found a value of $\alpha_v = 1$ for di-butyl phthalate evaporating in the presence of air, hydrogen or freon. However, as noted above in the discussion of theory, the value of α_v in free evaporation might still be less than one if the presence of the inert gas affected the activated state. Bradley and co-workers[241,242] later used the same technique to measure evaporation rates of other substances.

In summary, experiments have been performed which are in agreement with the predictions of equations (D-24) to (D-27).

3. Growth and Evaporation of Dislocation-Free Crystals

(a) *Non-Singular or "Roughened" Surfaces*

1. *Theory*

Frank[258] and Cabrera[38] have classified crystal bounding surfaces as: (i) *singular*, corresponding to a cusp in a Gibbs–Wulff plot, and usually to a

* LaMer and Gruen[529] have investigated the diffusion-controlled growth of dioctyl phthalate aerosols by absorption of toluene from atmospheres arising from master solutions of dioctyl phthalate and toluene. It is interesting to note that their results constitute the first direct experimental proof of the Gibbs–Thomson relationship $kT \ln (p/p_e) = 2\sigma\Omega/r$.

† Note that D and Δ are inversely proportional to p_{inert}.

low-index surface; (ii) *vicinal*, corresponding to a groove in a Gibbs–Wulff plot, and usually to a step structure in which low-index steps are separated by monomolecular "risers", and (iii) *non-singular*, corresponding to a smooth surface in a Gibbs–Wulff plot.* Early classifications of crystal growth[260] implied that non-singular surfaces would follow the same growth laws as liquids (see last section), because growth sites on the surface are not restricted. However, it is evidently possible that a surface which is non-singular at *equilibrium* with the vapor may be vicinal in the presence of a supersaturated vapor. The condition for this case is that the relaxation time for roughening of a low-index surface be long compared to the period for addition of a monomolecular layer to the surface. Nevertheless, one treatment of the formation of a kinetically vicinal surface[261] indicates that non-singular equilibrium surfaces will advance uniformly, like a liquid, when the supersaturation for growth is very small. Also, we will show later that vicinal surfaces remain vicinal up to the critical supersaturation for two-dimensional nucleation. Thus the equilibrium classification suffices in a consideration of crystal growth from the vapor, and it is concluded that solids which have non-singular equilibrium bounding surfaces should follow the kinetic growth and evaporation laws for liquids as developed in the last section.

2. *Experiment*

We know of no experimental results for the measurement of α_{v_i} or α_{c_i} on a surface which has been shown to be a non-singular equilibrium surface.

(b) *Singular and Vicinal Surfaces*

1. *Theory of Growth*

(a) *Clean crystals with monatomic vapor phase.* In the growth of a singular or vicinal surface, the surface structure envisaged by Kossel[2] and Stranski[3] (Fig. 28) is assumed. It was implied by Gibbs[68] and shown by Burton, Cabrera and Frank[260] (for the case of growth) and by Hirth and Pound[262] (for the case of evaporation) that the kinetics for such surfaces involve the steps: (a) adsorption or desorption of atoms at the surface, (b) surface diffusion,† and (c) transfer at kink positions in monatomic ledges on the crystal surface. If a perfect crystal with random surfaces is exposed to a supersaturated vapor, the ledges on {hkl} surfaces will grow out of the crystal and leave the crystal bounded by only low-index surfaces. Further

* Frank[259] has pointed out that if one plots the reciprocal of the surface energy versus orientation, the Gibbs–Wulff plot is simplified; typically a cusp becomes a polyhedron corner in a reciprocal plot, a groove becomes a polyhedron edge, and the spherical surfaces become planes.

† See discussion preceding equation (C-9). For solids it is most likely that the activational free energy for surface diffusion $\Delta G^{\star}_{sd} > kT$. Hence the adsorbed layer probably cannot be considered as a two-dimensional gas.

growth of the crystal is then controlled by two-dimensional nucleation of new monolayers upon the low-index surfaces, as was first noted by Gibbs.[68]

Two-dimensional nucleation has been recently treated by Burton and Cabrera,[263] by Kaischew[264] and by Hirth[144] following essentially the classical methods outlined in Section B and neglecting the statistical factor,

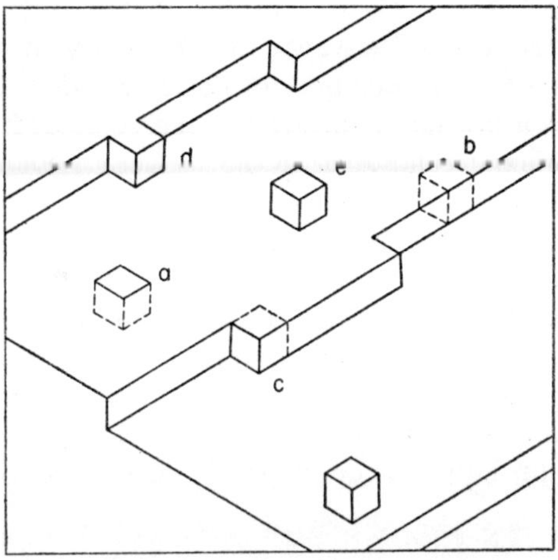

FIG. 28. Schematic view of metal–vapor interface depicting ledges and atoms in the following positions: *a*, in surface; *b*, in ledge; *c*, kink; *d*, at ledge; *e*, adsorbed on the surface.

equation (C-14). The concentration of single adsorbed atoms on a close-packed surface in metastable equilibrium with a supersaturated vapor is obtained* by equilibrating the gross vaporization flux $n_s \nu \exp(-\Delta G^{\star}_{\text{des}}/kT)$ with the gross condensation flux $p/(2\pi mkT)^{\frac{1}{2}}$, giving

$$n_s = [p/(2\pi mkT)^{\frac{1}{2}}]/\nu \exp(-\Delta G^{\star}_{\text{des}}/kT). \tag{D-28}$$

Consider the formation of a disc-shaped nucleus on the surface. The free energy of formation of the embryo ΔG^0, neglecting statistical contributions as developed in equation (C-14), will be

$$\Delta G^0 = 2\pi r\varepsilon + \pi r^2 h \Delta G_v \tag{D-29}$$

where, as before, $\Delta G_v = -(kT/\Omega) \ln(p/p_e)$ is the volume free energy change, r the radius of the disc, h the height of the monolayer and ε is the energy per unit length of monomolecular edge. Minimizing ΔG^0 with respect to r, one obtains the free energy of formation of the critical nucleus

$$\Delta G^{\star} = -\pi\varepsilon^2/h\Delta G_v. \tag{D-30}$$

* Compare with equation (D-6).

If one now supposes that the population of critical nuclei is equilibrated with that of single adsorbed atoms, the van't Hoff isotherm yields the classical value for the concentration of nuclei of critical size

$$n^{\star} = n_s \exp(-\Delta G^{\star}/kT). \tag{D-31}$$

Introducing the statistical factor (equation C-14) one obtains

$$n^{\star} = n_0 \exp(-\Delta G^{\star}/kT). \tag{D-32}$$

Including non-equilibrium corrections,[144] the nucleation rate is given by

$$J = Z\omega^{\star} n^{\star} \tag{D-33}$$

where Z is the Zeldovich non-equilibrium factor to account for depletion of the population of critical nuclei due to their growth or decomposition and $\omega^{\star}$ is the frequency with which these discs become supercritical. $\omega^{\star}$ is given by the product of the number $(n_s 2\pi r a)$ of adsorbed atoms in a position to join a critical nucleus times the frequency $[\bar{\omega} \exp(-\Delta G^{\star}_{\text{sd}}/kT)]$ with which such an adatom will jump to join the nucleus. Here $\bar{\omega} \simeq \nu$ is given by equation (D-5). Hirth[144] has shown that $Z = (\Delta G^{\star}/4\pi kTi^{\star 2})^{\frac{1}{2}} \simeq 10^{-2}$, where $i^{\star}$ is the number of atoms in the critical nucleus. Thus substituting in equation (D-33), the rate of two-dimensional nucleation (cf. equation C-44) becomes

$$J = (\Delta G^{\star}/4\pi kTi^{\star 2})^{\frac{1}{2}} 2\pi r^{\star} a \bar{\omega} n_s n_0 \exp(-\Delta G^{\star}_{\text{sd}}/kT) \exp(-\Delta G^{\star}/kT). \tag{D-34}$$

The pre-exponential in the last expression is $\sim e^{65}$, and the rate of nucleation is quite sensitive to the ΔG_v term in the exponential (see equation D-30). Accordingly the nucleation rate increases from a very small to a very large number in some narrow interval of ΔG_v. It is convenient to locate this interval at a value of ΔG_v for which $J = 1$ cm^{-2} sec^{-1} and $\ln J = 0$. Thus there is effectively a critical supersaturation ratio for an appreciable nucleation rate, and from equations (D-30) and (D-34)*

$$(p/p_e)_{\text{crit}} = \exp(\pi \varepsilon^2 \Omega / 65hk^2T^2). \tag{D-35}$$

Generally it is assumed (perhaps incorrectly) that $\varepsilon = h\sigma$ where σ is the specific solid–vapor interfacial free energy and hence

$$(p/p_e)_{\text{crit}} = \exp(\pi h \Omega \sigma^2 / 65k^2T^2). \tag{D-37}$$

Now on a low-index surface an adsorbed molecule will have a mean free path (see equation A-10) given by the Einstein relation

$$\bar{X} = (2D_s\tau_s)^{\frac{1}{2}} = [2a^2 \exp(\Delta G^{\star}_{\text{des}} - \Delta G^{\star}_{\text{sd}})/kT]^{\frac{1}{2}} \tag{D-38}$$

* The use of equation (D-31) instead of equation (D-32) would lead to the classical result

$$(p/p_e)_{\text{crit}} = \exp(\pi \varepsilon^2 \Omega / 50hk^2T^2). \tag{D-36}$$

If the spacing $\bar{\lambda}$ between monatomic ledges, which are sinks for adsorbed atoms, is small relative to $\bar{X}$, a situation which will obtain when the nucleation rate is rapid, almost all adsorbed atoms will reach sinks and the growth rate will be ideal. If $\bar{\lambda} \gg \bar{X}$, which occurs when the nucleation rate is small, very few adsorbed atoms will reach sinks and the growth rate will be effectively zero under such conditions. Thus the growth rate will go from zero to the ideal growth rate, and α_{c_i} will go from zero to unity at the critical supersaturation ratio given by equation (D-37). This effect has been described quantitatively by Becker and Doering[70] and by Burton and Cabrera[263] who used the classical equation (D-36) and is shown in Fig. 29. Equation (D-35) would of course predict a lower critical supersaturation ratio.

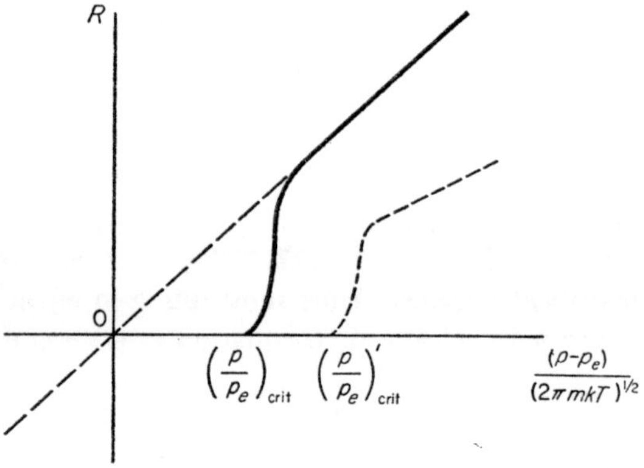

FIG. 29. Growth rate R versus supersaturation for a perfect crystal growing by nucleation of two-dimensional layers on low-index faces. The solid line is for a clean surface with no entropy constraint in growth. The dashed line is for a contaminated surface and/or a surface with an entropy constraint.

(b) *Entropy constraints.* Entropy constraints in the growth of the critical nucleus would have the effect of introducing a term δ into the frequency factor of equation (D-34) and thus

$$(p/p_e)'_{\text{crit}} = \exp\,(\pi h \Omega \sigma^2 / 65 \delta k^2 T^2). \qquad \text{(D-39)}$$

Thus entropy constraints would tend to cause $(p/p_e)'_{\text{crit}}$ to exceed $(p/p_e)_{\text{crit}}$. Also, once the critical supersaturation ratio had been exceeded, the δ effects could still lead to a value of $\alpha_c < 1$. The resulting expected growth rate in such a case is also shown in Fig. 29.

(c) *Surface roughening.* Cahn[261] has considered the possibility that a singular interface may become diffuse under a sufficient supersaturation. In this case the growth will change from a two-dimensional nucleation mechanism to uniform advance of a roughened interface. Now as shown in Fig. 29, the growth rate for the two-dimensional nucleation mechanism should become ideal for supersaturation ratios greater than $(p/p_e)_{\text{crit}}$. Hence the inception of a diffuse interface is important in its effect on the rate of crystal growth only if the interface becomes diffuse at a supersaturation ratio less than $(p/p_e)_{crit}$. The supersaturation ratio that must be attained for roughening[261] is

$$(p/p_e)_{diff} \geq \exp(\varepsilon^2/\sigma kT) \quad \text{(D-40)}$$

where the edge energy $\varepsilon = h^2\sigma/w$ and w is the width of the interface. Comparing equations (D-35) and (D-40), it is seen that the onset of a diffuse interface occurs at a supersaturation ratio less than $(p/p_e)_{crit}$ only if $\sigma > 2T$ (where σ is expressed in ergs/cm² and T in °K), a condition unlikely to obtain in crystal growth from the vapor. Hence the onset of diffuseness at the interface is unlikely to be important in affecting the rate of crystal growth, although an effect on the structure of the growing crystal is possible.

(d) *Incoherent crystal growth.* Finally, the structure of a growing crystal is considered. In developing equation (D-35) it was supposed that the interface between the two-dimensional nucleus and the low-index substrate was coherent. Sears[265] has developed an expression for the probability of formation of a two-dimensional nucleus which forms an incoherent interface with the substrate.* In such a case the free energy of formation of the nucleus is (compare with equation D-29)

$$\Delta G^\star_\perp = 2\pi r^\star \varepsilon + \pi r^2(h\Delta G_v + \sigma_\perp) \quad \text{(D-41)}$$

where $\sigma_\perp$ is the interfacial free energy of the incoherent interface. The frequency factors in the nucleation expression are the same as before. Thus the ratio of probabilities for the two modes of nucleation is from equation (D-34)

$$J_\perp/J = \exp[\beta\Delta G^\star/(1+\beta)kT] \quad \text{(D-42)}$$

where

$$\beta = \sigma_\perp/h\Delta G_v. \quad \text{(D-43)}$$

At the critical supersaturation, $\beta \cong -40\sigma_\perp kT/\pi h^2\sigma^2$, and

$$(J_\perp/J)_{crit} = \exp[-40\beta/(1+\beta)]. \quad \text{(D-44)}$$

Taking[267] $\sigma = 1000$ ergs/cm², $T = 1000$°K, $h^2 = 10^{-15}$ cm², $\sigma_\perp = 20$ ergs/cm² for a stacking fault, $\sigma_\perp = 200$ ergs/cm² for a highly disordered incoherent interface, one gets $\beta \cong 0.04$, $(J_\perp/J)_{crit} \cong 0.25$ for a stacking fault and $\beta \cong 0.4$, $(J_\perp/J)_{crit} \cong 10^{-8}$ for a disordered interface. Hence in growth from the vapor by a two-dimensional nucleation mechanism, there is a high probability that some stacking faults will be formed in the growing crystal but very low probability that an incoherent interface will be formed. Of course these conclusions hold only when condition (D-40) is not fulfilled.

In summary, Fig. 29 shows the expected kinetic behavior of a crystal

* Such incoherent interfaces are bounded by dislocations if they terminate within a crystallite.[266] Nevertheless, this type of growth is considered in this section because the dislocations form after growth has begun and are not required for growth to take place.

growing by a two-dimensional nucleation mechanism. Crystals that grow by such a mechanism are likely to contain grown-in stacking faults.

2. *Theory of Evaporation*

(a) *Nucleation of a hole at the surface.* Nucleation is not required to supply evaporation steps because they can arise at the edges of a clean singular surface, as discussed in the next part of the present section. However, in the case of contaminated crystals the edges may be protected and nucleation on a perfect surface may be important. Now even though a perfect crystal surface may be exposed to an undersaturation less than the critical without nucleation of a disc-shaped hole, equilibrium between the undersaturated adatom population and vacancies at the surface can be established as long as the relaxation time for surface diffusion (equation A-8) is short compared with a typical observation time. The treatment of a two-level model of the surface,[268] which should be adequate as long as $n_s \ll (1/a)^2$, the concentration of atoms in the surface, predicts that n_{h_e}, the equilibrium concentration of surface vacancies is given by

$$n_{h_e} = n_{s_e}. \tag{D-45}$$

Under non-equilibrium steady-state conditions the rate formation of vacancies is given by the product of $(1/a)^2$, and the rate of formation per site, $\bar{\omega} \exp(-\Delta G_f/kT)$. The rate of annihilation is given by $n_h n_s a^2$, the concentration of adatoms adjoined by vacancies, times the frequency with which an adjoining adatom will jump into the vacant site, $\bar{\omega} \exp(-\Delta G^\star_{sd}/kT)$. All terms but n_s and n_h are independent of undersaturation, and upon equating the rates of formation and annihilation one obtains

$$n_s n_h = \text{constant} = n_{s_e} n_{h_e}$$

or

$$n_h = n_{h_e} n_{s_e}/n_s = n_{s_e}{}^2/n_s. \tag{D-46}$$

The free energy of formation of a disc-shaped embryo is given by equation (D-29) with

$$\Delta G_v = -(kT/\Omega)\ln(n_h/n_{h_e}) = -(kT/\Omega)\ln(n_{s_e}/n_s) = (kT/\Omega)\ln(p/p_e) \tag{D-47}$$

according to equations (C-1) and (D-46). The free energy of formation of the critical nuclei is expressed by equation (D-30) and their equilibrium concentration by (D-32). The nucleation rate is given by equation (D-33) and Z is the same except that $i^\star$ now refers to the number of vacant sites in the critical nucleus. This leads to

$$n^\star = n_0 \exp(-\Delta G^\star/kT) \tag{D-48}$$

where the statistical factor[65] (n_0/n_h) has been introduced.

The critical nuclei in the surface may become supercritical by one of three possible atomic processes: (a) an atom at the edge of a critical nucleus may evaporate directly into the vapor phase, (b) it may dissociate to an adsorbed position on the low-index surface and subsequently desorb into the vapor, (c) a vacancy adjacent to a critical nucleus may jump to join the nucleus. Analogously to the case of nucleation in crystal growth,[144] the direct evaporation process can be ruled out by consideration of the principle of microscopic reversibility. Also by this principle the frequency with which an atom dissociates from a ledge position on a disc will equal the reverse frequency if the metastable equilibrium population of adatoms exists there.[144] Thus the frequency factor for such a case is given by the product of the number $2\pi r^\star a n_{s_e}$ of adatoms adjoining a critical nucleus and their jump frequency to join it

$$\omega^\star = (2\pi r^\star a n_{s_e} \nu) \exp(-\Delta G^\star_{\text{sd}}/kT).^* \tag{D-49}$$

The frequency factor in the case of an adjacent vacancy jumping to join the nucleus will be

$$\omega^\star = 2\pi r^\star a n_h \nu \exp(-\Delta G^\star_{\text{hd}}/kT) \tag{D-50}$$

where $\Delta G^\star_{\text{hd}}$ is the activational free energy for the jump of a vacancy near the surface. It is likely that $\Delta G^\star_{\text{hd}} > \Delta G^\star_{\text{sd}}$ and hence the frequency factor given by equation (D-49) should be larger at low undersaturations where $n_s \simeq n_{s_e}$. However, the frequency factor given by equation (D-50) should be larger at high undersaturations where $n_{s_e} > n_s$. Substituting into equation (D-33) and using equation (D-50) for $\omega^\star$, in the usual case of high undersaturation one obtains:

$$\begin{aligned} J = 2\pi r^\star a n_h n_0 \nu (\Delta G^\star/4\pi kT i^{\star 2})^{\frac{1}{2}} \exp(\Delta G^\star_{\text{hd}}/kT) \exp(-\Delta G^\star/kT) \\ = \exp B \exp(-\Delta G^\star/kT). \end{aligned} \tag{D-51}$$

The critical value of $\Delta G^\star$ for appreciable nucleation is obtained with $J =$ one nucleus/cm^2 sec or with $\ln J = 0$ and thus

$$(p/p_e)_{\text{crit}} = \exp[-\pi\varepsilon^2\Omega/hB(kT)^2]. \tag{D-52}$$

Analogously to the discussion following equation (D-38), it may be shown that the evaporation rate by the disc-nucleation mechanism will be essentially zero for $(p/p_e) > (p/p_e)_{\text{crit}}$ and should be approximately the ideal rate for $(p/p_e) < (p/p_e)_{\text{crit}}$, where $(p/p_e)_{\text{crit}}$ is given by equation (D-52).

(b) *Molecular kinetics.* The theory of evaporation of crystals has been treated by Knacke *et al.*[269] and by Hirth and Pound,[41,262] following ideas developed by Gibbs,[68] Kossel,[2] Stranski[3] and Burton *et al.*[260] The

* Actually it might be expected that n_{s_e} should represent the hypothetical adatom population in equilibrium with the disc and should contain a term $\exp(-\sigma\Omega/rkT)$ to account for the effect of capillarity on the chemical potential of an atom at the disc. However, in the present treatment this capillarity correction appears in the factor Z (see ref. 144).

distinction between evaporation kinetics and growth kinetics lies in the role of crystal edges. Unlike their role in growth,* crystal edges are favored sites for monomolecular ledge formation. Hence not only are vicinal surfaces maintained in crystal evaporation, but even a perfect crystal which is bounded by singular surfaces will exhibit vicinal surfaces due to the formation of ledges at crystal edges and the subsequent motion of these ledges onto the singular surface. The kinetics of evaporation then involve: dissociation of molecules from kink sites to positions at ledges (see Fig. 28), diffusion along the ledge, dissociation from the ledge to an adsorbed position, diffusion of the adsorbed molecule, and desorption to the vapor. The diffusion equations for diffusion along the ledge and on the surface both involve moving boundary conditions. However, these may be neglected[260] if, as is almost always the case, $D_L/\bar{Y} > v_y$ and $D_s/\bar{X} > v_x$, where D_L and D_s are the diffusion coefficients for ledge and surface, $\bar{Y}$ and $\bar{X}$ are the mean free paths of diffusion on a ledge and on a surface, and v_y and v_x are the velocities of kinks and ledges, respectively.

As will be shown in the next section where imperfect crystals containing screw dislocations are treated, the curvature of ledges in steady-state evaporation is in general sufficiently small that the problem of motion of a train of ledges can be approximated as one-dimensional. One can write for the net steady-state flux of molecules leaving kink positions, in atoms/cm² sec,

$$J_1 = (2n_k\nu/\lambda)\exp(-\Delta H^{\star}_{k-1}/kT) - (2an'_L n_k \nu \delta_{k-1}/\lambda)\exp(-\Delta H^{\star}_{ld}/kT) \tag{D-53}$$

where n_k is the number of kinks per cm, λ the interledge spacing, $\Delta H^{\star}_{k-1}$ the enthalpy of activation to move from a kink to a position at a ledge, n'_L the concentration of molecules at a ledge and immediately adjacent to a kink, δ_{k-1} the "free-angle ratio" for kink–ledge kinetics,† and $\Delta H^{\star}_{ld}$ is the enthalpy of activation for diffusion at ledges. A ledge–diffusion flux will be set up away from kink sites, and the divergence of this flux will equal the net flux of molecules from ledge positions to surface positions. Thus

$$D_L(\partial^2 n_L/\partial y^2) = 2n_L\nu\exp(-\Delta H^{\star}_{l-ad}/kT) - 2n'_s\delta_{l-ad}a\nu\exp(-\Delta H^{\star}_{sd}/kT) \tag{D-54}$$

where n_L is the concentration of molecules at a ledge, $\Delta H^{\star}_{l-ad}$ is the activation energy to dissociate from a ledge to an adsorbed position, n'_s is the value of n_s

* For highly polarized surfaces, edges are slightly favored as sites for nucleation[270,271] in condensation and thus are protected to some extent in evaporation. The above discussion holds for metals and homopolar crystals. Available evidence, however, indicates that most ionic crystals behave as intermediate between homopolar and heteropolar behavior. In general (see experimental section) edges in ionic crystals appear to act as sources for evaporation ledges.

† cf. equation (D-13). In each kinetic step, rotational degrees of freedom are considered to be hindered in the activated state, giving rise to δ terms. Otherwise, if rotational degrees of freedom were activated, the rotational terms would appear in a standard free energy term in lieu of $\Delta H^{\star}_{k-1}$ in equation (D-53) and would not affect α_v, i.e. would not cause deviations from the ideal evaporation equation (A-2).

at a ledge,* $\delta_{\text{l-ad}}$ is the "free-angle ratio" for the ledge-adsorbed kinetics, and $\Delta H^{\star}_{\text{sd}}$ is the activational enthalpy for surface diffusion. The boundary conditions are $n_L = n'_L$ at kinks and $(\partial n_L/\partial y) = 0$ midway between two kinks. Solving for n_L from (D-54) and integrating to find the average value $\bar{n}_L$ one obtains for the net steady-state flux of molecules at ledges

$$\begin{aligned} J_2 &= (2\bar{n}_L \nu/\lambda) \exp(-\Delta H_{\text{l-ad}}/kT) - (2n'_s a\nu\delta_{\text{l-ad}}/\lambda) \exp(-\Delta H^{\star}_{\text{sd}}/kT) \\ &= (2/\lambda)(\sqrt{2}\,\bar{Y}/\Psi) \tanh(\Psi/\sqrt{2}\,\bar{Y})[n'_L\nu \exp(-\Delta H^{\star}_{\text{l-ad}}/kT) \\ &\qquad - n'_s a\nu\delta_{\text{l-ad}} \exp(-\Delta H^{\star}_{\text{sd}}/kT)] \\ &= (2/\lambda)\alpha_\Psi[n'_L\nu \exp(-\Delta H^{\star}_{\text{l-ad}}/kT) - n'_s a\nu\delta_{\text{l-ad}} \exp(-\Delta H^{\star}_{\text{sd}}/kT)] \end{aligned} \tag{D-55}$$

where Ψ is the spacing between kinks.

Similarly, a diffusion flux will be set up from ledges onto low-index surfaces, the divergence of which equals the net vaporization flux. Thus

$$D_s(\partial^2 n_s/\partial x^2) = n_s\nu \exp(-\Delta H^{\star}_{\text{des}}/kT) - \delta_{\text{ad}}p/(2\pi mkT)^{\frac{1}{2}} \tag{D-56}$$

where δ_{ad} is the "free-angle" term for adsorption from the vapor. The boundary conditions on (D-56) are $n_s = n'_s$ adjacent to a ledge and $(\partial n_s/\partial x) = 0$ midway between ledges. Solving for n_s from (D-56) and integrating to obtain $\bar{n}_s$, one gets for the net vaporization flux from low-index planes[262]

$$\begin{aligned} J_3 &= [(\sqrt{2}\bar{X}/\lambda) \tanh(\lambda/\sqrt{2}\bar{X})][n'_s\nu \exp(-\Delta H^{\star}_{\text{des}}/kT) - \delta_{\text{ad}}p/(2\pi mkT)^{\frac{1}{2}}] \\ &= \alpha_\lambda[n'_s\nu \exp(-\Delta H^{\star}_{\text{des}}/kT) - \delta_{\text{ad}}p/(2\pi mkT)^{\frac{1}{2}}]. \end{aligned} \tag{D-57}$$

Now at steady state $J_1 = J_2 = J_3$, and so by solving equations (D-53), (D-55) and (D-57) simultaneously and eliminating n'_s and n'_L, one finally obtains the general evaporation flux equation, cf. reference (272),

$$J_v = \frac{(\alpha_\lambda\nu/\delta_{\text{k-l}}\delta_{\text{l-ad}}a^2) \exp - [(\Delta H^{\star}_{\text{k-l}} + \Delta H^{\star}_{\text{l-ad}} + \Delta H^{\star}_{\text{des}} - \Delta H^{\star}_{\text{ld}} - \Delta H^{\star}_{\text{sd}})/kT] - \alpha_\lambda\delta_{\text{ad}}p/(2\pi mkT)^{\frac{1}{2}}}{1 + (\lambda\alpha_\lambda a/\alpha_\Psi\delta_{\text{l-ad}}\bar{X}^2)[1 + (2\alpha_\Psi a/n_{\text{k}}\delta_{\text{k-l}}\bar{Y}^2)]} \tag{D-58}$$

This equation can be reduced for specific cases; some likely examples follow.

(i) *Clean crystalline surface with monatomic vapor phase.* In such a case, $\delta_{\text{k-l}} = \delta_{\text{l-ad}} = \delta_{\text{ad}} = 1$. Also it is likely that $\Delta H^{\star}_{\text{kl}}$ equals $\Delta H_{\text{k-l}}$, the equilibrium enthalpy term, plus $\Delta H^{\star}_{\text{ld}}$. Similarly $\Delta H^{\star}_{\text{l-ad}} \simeq \Delta H_{\text{l-ad}} + \Delta H^{\star}_{\text{sd}}$ and $\Delta H^{\star}_{\text{des}} \simeq \Delta H_{\text{des}}$. The denominator in equation (D-58) becomes unity, and noting that

$$(\nu/\delta_{\text{k-l}}\delta_{\text{l-ad}}a^2) \exp[-(\Delta H_{\text{k-l}} + \Delta H_{\text{l-ad}} + \Delta H_{\text{des}})/kT] = \delta_{\text{ad}}p_e/(2\pi mkT)^{\frac{1}{2}}, \tag{D-59}$$

* It is supposed that n'_s is not a function of y. This will be so if $D_s > D_L$, a condition which generally obtains. If $D_L > D_s$, $n_L = n_{L_e}$ and $n_s \simeq n_{s_e}$,[262] so the above supposition is self-consistent.

one obtains for equation (D-58)

$$J_v = \alpha_\lambda (p - p_e)/(2\pi mkT)^{\frac{1}{2}} \tag{D-60}$$

where*

$$\alpha_{v_6} = \alpha_\lambda = (\sqrt{2}\bar{X}/\lambda) \tanh (\lambda/\sqrt{2}\bar{X}). \tag{D-61}$$

This is the expression derived[262] directly for evaporation in such a case. It is noted that[262] ledges are perfect sources of adatoms, i.e. $n_s = n_{s_e}$, as a consequence of the above conditions.

(ii) *Entropy control at a ledge.* In such a case $\delta^{\prime}_{1-\mathrm{ad}} < 1$. However, it is likely that the above approximation may be taken for the pertinent enthalpies of activation. Here the last term in the denominator is unity while the second term should be relatively large. Accordingly one obtains

$$J_v = \frac{\alpha_\lambda \delta_{\mathrm{ad}}(p - p_e)/(2\pi mkT)^{\frac{1}{2}}}{[1 + (\lambda \alpha_\lambda a/\alpha_{\Upsilon'} \delta_{1-\mathrm{ad}} \bar{X}^2)]} \tag{D-62}$$

or*

$$\alpha_{v_7} = \alpha_\lambda \delta_{\mathrm{ad}} f(\delta_{1-\mathrm{ad}})/[\alpha_\lambda + f(\delta_{1-\mathrm{ad}})] \tag{D-63}$$

where $f(\delta_{1-\mathrm{ad}}) = (\delta_{1-\mathrm{ad}} \bar{X}^2 \alpha_{\Upsilon'}/\lambda a)$. Again, this is the expression which may be derived[41] directly for such a case.

(iii) *High activation energy for dissociation from kink.* If large activation energies are involved in evaporation, they will probably be present in the kink-adsorbed stage. In such a case $\alpha_{\Upsilon'} < 1$, $n_{\mathrm{K}} \ll 1/a$ and equation (D-58) reduces to

$$J_v \cong (2n_{\mathrm{K}}\nu/\lambda) \exp [-(\Delta H^{\star}_{\mathrm{K}-1} + \Delta H^{\star}_{1-\mathrm{ad}})/kT] \tag{D-64}$$

which, again, is the expected result that one would obtain directly.

Other examples may be derived from equation (D-58). In general, unless the conditions for equation (D-64) obtain, the last term in the denominator can be neglected. In cases where diffusion in the vapor phase is important these kinetics must also be considered.[41] However, this consideration is likely to be more important in equilibrium vapor-pressure measurements than in free-evaporation experiments, because in the latter situation high vacua are generally used to speed diffusion in the vapor phase. The corrections necessary when diffusion in the vapor is important have been treated in Part 2 of this section and elsewhere.[41] Geometry effects must also be considered when a surface is not smooth. For example the true evaporation area will exceed the apparent area, and sharp crevices will act as quasi-Knudsen cells[273] to give fluxes from those areas which approach the equilibrium flux. Finally, a surface temperature correction may be required (cf. discussion in Part 2 of this section and the work of Littlewood and Rideal[239]). However, it is noted that self-heating or cooling effects due to the latent heat of vaporization are much smaller for solids than for liquids because of the lower fluxes at solid surfaces. Accordingly in metals of high thermal conductivity, temperature gradients are unlikely. Such effects are likely only in molecular substances of low thermal conductivity.

* As will be shown in Section E, there are equivalent α_{c_i} values $\alpha_{c_6} = \alpha_{v_6}$ and $\alpha_{c_7} = \alpha_{v_7}$.

(c) *Ledge kinetics.* Given equation (D-58), one must know the ledge spacing on a surface in order to describe the evaporation kinetics. Whenever surface-diffusion control obtains, as in equations (D-60) and (D-61), the ledge spacing will not attain a steady state but will approach an asymptotic state.[262] The velocity of a ledge is given, from equation (D-56), by

$$v = -(2D_s\Omega/h)(\partial n_s/\partial x)_{x=0} \quad \text{(D-65)}$$

$$= [\sqrt{2}D_s\Omega/\bar{X}h\nu \exp(-\Delta H^{\star}_{\text{des}}/kT)] \tanh(\lambda/\sqrt{2}\bar{X})[(p-p_e)/(2\pi mkT)]^{\frac{1}{2}} \quad \text{(D-66)}$$

Hence a train of ledges will originate at an edge and accelerate on the crystal surface as the ledge spacing increases. Thus λ and v will increase at a decreasing rate until $\tanh(\lambda/\sqrt{2}\bar{X}) \simeq 1$, when the asymptotic state is nearly attained. This occurs for $(\lambda/\sqrt{2}\bar{X}) \simeq 3$ and by equation (D-61) leads to an evaporation coefficient for a large perfect crystal bounded by low-index places of

$$\alpha_{v_0} = \sqrt{2}\bar{X}/\lambda \simeq \sqrt{2}\bar{X}/\lambda_0 = \tfrac{1}{3}. \quad \text{(D-67)}$$

For any case in which surface diffusion enters the kinetics, as in equation (D-63), the ledge spacing should approach the asymptotic value at $\lambda_0/\sqrt{2}\bar{X} = 3$. However, except for the case of equation (D-61), other effects, such as δ effects, are likely to overshadow α_λ in controlling evaporation *kinetics*. Nevertheless equation (D-66) will still control the evaporation *topography*.

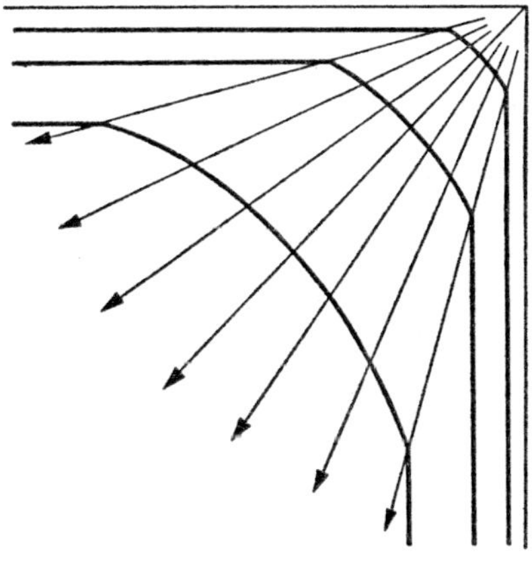

FIG. 30. Idealized cross section of a long square prism initially bounded by {100} planes as a function of evaporation time.

Evidently, because of the non-steady-state nature of the kinetics, the high-index planes originating at crystal corners will after a long period of time cover the crystal, leading to a shape for which the ledge spacing $\lambda \ll \bar{X}$ and $\alpha_\lambda \simeq 1$. A typical case is illustrated in Fig. 30. Clearly, for a crystal initially bounded by high-index planes, λ is *initially* small and hence throughout evaporation $\lambda \ll \bar{X}$ and $\alpha_\lambda \simeq 1$.

Frank,[274] following a mathematical treatment developed by Lighthill and Whitham[275] to describe traffic flow on highways, has developed a quantitative means of predicting the evaporation profile as shown in Fig. 30.* If the local evaporation rate $J(\lambda)$ is a function of λ only, then points of a given orientation (i.e. of a given ledge spacing) will move in a straight line in space. Furthermore, if one constructs a polar diagram of $[1/J(\lambda)]$ vs. orientation (i.e. λ), the straight-line space trajectory will be parallel to a normal to the polar diagram at the corresponding orientation. Thus if one knows $J(\lambda)$, as by equation (D-58), one can draw a predicted evaporation profile, Fig. 30. This theory predicts that if, as is the case for equation (D-66), $\partial^2|J_v|/\partial(1/\lambda)^2$ is negative: (a) the evaporation profile will appear as in Fig. 30, and (b) a "bunch" of ledges (corresponding to a macroscopic ledge) will diverge or flatten out with time. We note the reservation in applying this theory to evaporation that *near a ledge source*, such as a crystal edge or a macroscopic ledge, λ is changing very rapidly with time due to ledge acceleration and hence J_v is a function of both λ and $(\partial\lambda/\partial x)$. This violates the condition that J_v is a function of λ only. Therefore we feel that, while Frank's model will apply over most of the evaporation surface, the trajectories near ledge sources may not be straight lines.

Finally, it is noted that Frank's theory, whether or not the above reservation applies, predicts that macroscopic ledges are *unstable*† unless an impurity adsorbs onto a crystal in a time-dependent manner. Thus the presence of macroscopic ledges in evaporation or growth indicates that impurity adsorption is likely to be affecting the kinetics.

In summary for clean perfect crystals: evaporation will proceed at any finite undersaturation because crystal edges act as ledge sources; the asymptotic limiting evaporation coefficient for a large low-index crystal surface is, in the absence of entropy constraints, $\alpha_{v_6} \simeq \frac{1}{3}$; for a high-index crystal surface, $\alpha_{v_6} \simeq 1$; entropy constraints or high enthalpies of activation lead to $\alpha_{v_7} \ll \frac{1}{3}$, in which case surface diffusion no longer markedly affects the value of α_v but does affect the topography.

3. *Experiment*

(a) *Growth.* The number of experiments on growth and evaporation of perfect‡ crystals is quite limited. However, it will be shown in the next section that the kinetic laws for imperfect crystals are identical in many cases

* The theorem is general and applies to growth profiles as well. For our purposes the evaporation application is the most useful. Also it is noted that Vermilyea and Cabrera[276] and Chernov[277] have developed treatments similar to Frank's. Chernov approached the problem in the same manner as Frank but did not solve it completely. Vermilyea and Cabrera's solution is not as general as Frank's but they do consider the possible effects of impurities on growth kinetics in somewhat greater detail.

† This is true only for imperfection-free crystals. The presence of stacking faults may also lead to macroscopic ledge formation as discussed in the next section.

‡ Here we use "perfect" in the limited meaning that the crystals are free from dislocations. Thus we exclude from this definition point defects, impurity atoms, etc.

with those for perfect crystals, justifying the above extensive theoretical treatment. The principal experiments confirming the expected growth and evaporation behavior of perfect crystals were not performed on perfect crystals at all.* Some of these experiments were on whiskers which had perfect prismatic faces but an axial dislocation which leads to an imperfection at the whisker tip. Others were on platelets which had perfect planes but grids of dislocations that intersect lateral faces. Consideration of the growth of the imperfect faces is deferred to the next section. Here we will consider only the perfect faces of whiskers and platelets.†

Volmer and Estermann[230,270] found that in a vacuum of $\sim 10^{-5}$ mm Hg and a supersaturation of $\sim 10^3$ mercury crystals grew on a glass substrate at $-63°$C in the form of thin platelets with the large faces parallel to low-index (100) rhombohedral planes. The growth rate normal to the (100) planes was ~ 0.1 of the ideal rate calculated from the vapor impingement flux and equation (A-2). On the other hand, the growth rate of the bounding planes at the edges was 1000 times the ideal rate. Therefore they suggested that growth took place by adsorption, surface diffusion on (100) planes and accretion at edges, but that little accretion (and hence two-dimensional nucleation) occurred on the (100) planes. Sears[16,282] confirmed Volmer and Estermann's findings under conditions which were similar except that his vacuum was $\sim 10^{-6}$ mm Hg. Sears platelets were about 0.3 mm on a side and the calculated $\bar{X}$ was 25 microns. Hence over most of the (100) surface, equilibration of adatom concentration with the vapor obtained and the adatom supersaturation ratio was equal to 10^3. Now from equation (D-36), taking $\sigma = 500$ ergs/cm^2, one calculates $(p/p_e)_{\text{crit}} = 10^6$ for two-dimensional nucleation. Therefore the observation that little two-dimensional nucleation occurred on (100) faces at $(p/p_e) = 10^3$ is consistent with the prediction of equation (D-36) or (D-37). However, one notes that in a vacuum of 10^{-5} or 10^{-6} mm Hg, impurity adsorption may have occurred on the (100) planes and hence equation (D-68) may be more appropriate for description of the results.

Sears[16,283] also observed the growth of mercury whiskers from the vapor under conditions identical with the above except that (p/p_e) was 10^2 to 10^3. Since whiskers grew, there must have been no two dimensional nucleation on the prism faces of the whiskers. Again, this is consistent with theory, because (p/p_e) was about 10^2 while for two-dimensional nucleation $(p/p_e)_{\text{crit}}$ is 10^6.

Sears hypothesized that whisker growth should occur only if $(p/p_e) < (p/p_e)_{\text{crit}}$ because, as noted in the preceding paragraph, above the critical

* It is well established[258] that ordinary crystal faces contain dislocations. Exceptions are some whisker[278,279] and platelet[280] faces. To the best of our knowledge the only case in which a bulk crystal was grown free of dislocations was Dash's experiment[281] on pulling silicon crystals from the melt.

† It is surmised that the platelets and whiskers discussed here had perfect faces, because their growth rates were consistent with a model requiring perfect basal or prismatic faces. This, again, is mainly an argument of self-consistency.

supersaturation two-dimensional nucleation should be rapid and equiaxed crystallites should form rather than whiskers. He measured(169) the supersaturation above which whisker growth was not observed for several materials as shown in Table 9.

TABLE 9

CRITICAL SUPERSATURATION RATIO CALCULATED FROM EQUATION (D-36) AND OBSERVED SUPERSATURATIONS ABOVE WHICH WHISKER GROWTH WAS NOT OBSERVED FOR THE MATERIALS NOTED

Vacuum $\sim 10^{-6}$ mm Hg, glass substrate.

Substance	Calculated $(p/p_e)_{crit}$	Observed $(p/p_e)_{crit}$	T substrate	σ ergs/cm^2
Cd	14	$\sim$20	250°C	650[284]
Zn	6	$\sim$ 3	350°C	750[284]
Ag	5	$\sim$10	850°C	1140[285]
CdS	$\gtrsim$ 4	$\sim$ 2	800°C	—

Considering possible impurity-adsorption effects and the marked sensitivity of the calculated value of $(p/p_e)_{crit}$ to σ, the agreement of data in Table 9 with equation (D-36) (or D-37) is very good.

Samelson[286] performed an experiment in which the transition from whisker growth to equiaxed crystal growth of ZnS occurred at a (p/p_e) of 3 to 4 as compared with a calculated $(p/p_e)_{crit}$ of about 3. This work was carried out in a vacuum of 10^{-4} mm Hg at 1045 to 1151°C. Also Morelock[287] grew whiskers of chromium, nickel, iron, copper and gold, all at supersaturation ratios $(p/p_e) < 2$. He roughly calculated from equation (D-36) that $(p/p_e)_{crit}$ was greater than 2 in each case. Morelock examined the shapes of the whiskers and tested their strengths and found that their behavior was consistent with the hypothesis that the prism faces were perfect. Thus his results are also consistent with equation (D-36) (or D-37).

Price[288] studied the growth of cadmium in a cell in which the supersaturation of cadmium was established by diffusion through argon at $\sim$250°C and $\sim$1 atmosphere. He found that whiskers and platelets formed when $(p/p_e) \lesssim 3$ but that platelets thickened, indicating the onset of two-dimensional nucleation, when $(p/p_e) \gtrsim 3$. Thus Price's work indicates that $(p/p_e)_{crit} \sim 3$, in disagreement with Sears and with equation (D-36) (or D-37). Impurity adsorption from the "inert" argon could have led to Price's low observed critical supersaturation.* Price[289] also viewed the cadmium

* Price[290] points out that an alternative explanation is that the value of σ cited in Table 9 and used in equation (D-36) is too high. A lower value of σ (525 ergs/cm^2) would lead to a calculated $(p/p_e)_{crit}$ of $\sim$3. Further work, particularly high-vacuum work, appears to be needed to resolve this question.

platelets by transmission electron microscopy and confirmed that many of the platelets were perfect.

Also, De Vries and Sears[291] grew Al_2O_3 whiskers at 1800°C under conditions such that the temperature of the whisker tip decreased as the whisker grew. Hence the supersaturation ratio (p/p_e) on the prism faces near the tip increased as the whisker grew. At a length such that (p/p_e) at the tip reached $(p/p_e)_{crit}$ they observed that the tip started thickening and that layers grew from the tip towards the whisker base. Again this is in qualitative agreement with equation (D-36) or (D-37).

(b) *Evaporation.* Sears[292,293] also investigated the evaporation of perfect crystals. He grew platelets of para-toluidine at about 50°C and at low supersaturation by what is presumed to be the mechanism discussed above. Having thus obtained perfect (001) faces on the platelets, he exposed a small region of the (001) face to an undersaturated jet of vapor while the remainder of the crystal surface was in contact with saturated vapor. He found[292] that evaporation from the area of unsaturation took place only when $0.35 < (p/p_e) < 0.48$, and that no evaporation occurred when $0.48 < (p/p_e) < 1$. When the edges of a perfect crystal are protected, evaporation should proceed by two-dimensional nucleation. Taking $(p/p_e)_{crit} \sim 0.48$, one finds agreement with equation (D-52) if $\sigma = 53$ ergs/cm^2 for solid *p*-toluidine. This is a reasonable value in view of the measured σ of 34.5 ergs/cm^2 reported[294] for liquid *p*-toluidine. On the other hand[293] when the jet was directed near, but not at, a crystal edge, evaporation took place at an undersaturation ratio of $1 > (p/p_e) > 0.975$. This is an agreement with the prediction[262] mentioned above that crystal edges serve as sources for evaporation ledges even at very small undersaturations. More recently, Hudson and Sears[295] studied the evaporation of the edges of perfect prism faces of zinc whiskers at 418°C. They found that evaporation occurred at $1 > (p/p_e) > 0.99999$. Again this agrees with the prediction[262] that crystal edges serve as ledge sources at very small undersaturations.

(c) *Crystal topography and structure.* Frank's ideas on crystal topography[274] and the prediction of growth trajectories as a function of time have not been tested in the case of crystal–vapor kinetics. However, crystal dissolution, which follows the same general surface kinetics as evaporation but differs in detail, has received some study. Here the time dependence of ledge velocity, which was discussed above, should be minimized. Ives and Frank[296,297] have shown that Frank's theory[274] does exactly describe the macroscopic dissolution trajectories found by Batterman[298] for germanium and by Ives[297] for LiF.*

Although perfectly oriented films of metal with a low-index surface have

* This does not imply that the orientation trajectory can always be predicted by Frank's model. A case in which the ledge velocity was time-dependent while the dissolution trajectories were *not* straight lines has been reported by Ives and Hirth.[299] Frank, of course, did not intend that his theory should apply unless dissolution rate is a function only of orientation.

been grown from the vapor,[300,301] stacking faults have been observed in silver, gold and copper single-crystal foils which were similarly grown at presumably high supersaturation.[302–304,189] Some of these faults may have arisen from vacancy condensation or growth mistakes,[302,303] but their presence may also be interpreted as confirming Sears'[265] view on incoherent nucleation. Similarly the finding[304] that aluminum which was grown from the vapor did not contain stacking faults is consistent with Sears' view because of the high stacking-fault energy of aluminum. A more compelling confirmation of Sears'[265] prediction is found in the work of Sloope and Tiller.[197] They grew monocrystalline films of silver on NaCl substrates and observed that the stacking-fault density increased with either increasing deposition rate at constant substrate temperature or decreasing substrate temperature at constant deposition rate. Both of these findings indicate an increase in stacking-fault probability with increasing supersaturation as required by equation (D-44).

In summary, general agreement between theoretical expectation, as summarized in Fig. 29, and experiment has been found. The distinction between the roles of crystal edges in evaporation and in growth of perfect crystals has been demonstrated. Nevertheless, although critical supersaturations for two-dimensional nucleation have been found, the curves in Fig. 29 have not been reproduced in their entirety. Also, theories of macroscopic growth or evaporation trajectories have not been tested. Hence further experiments on these points appear to be desirable.

4. Effects of Impurities on the Growth and Evaporation of Perfect Crystals

(a) *Crystals with Non-Singular Surfaces*

Impurity effects in the growth and evaporation of such crystals would be identical with those on growth and evaporation of liquids as treated above.

(b) *Singular or Vicinal Surfaces*

1. *Theory of Crystal Growth*

Surface contamination would have an effect similar to that of the entropy constraints treated in Section (D-3) on crystal growth. Impurity adsorbate could affect the two-dimensional nucleation process in several ways as shown in Fig. 31. Sears[300] considered examples 2 and 3 and concluded that example 3 is unlikely. The detailed kinetics will differ for each case, but by considering the relative free energies of formation of the nuclei one can predict that the nucleation rate is likely to be enhanced by adsorbate only in cases 5 and 6. We feel that example 6, which was suggested by Stranski[305]

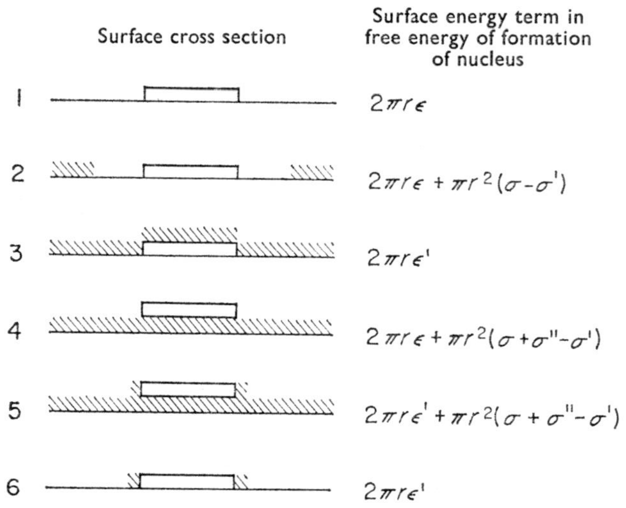

FIG. 31. Possible mechanism of effect of impurity on two-dimensional nucleation. Surface cross section of critical nuclei are shown. The hatching represents impurity adsorbate. $\varepsilon \simeq a\sigma$ is edge energy of clean nucleus; $\varepsilon' \simeq a\sigma'$ is edge energy of nucleus with impurity adsorbed at edge; σ'' is surface energy of interface between nucleus and substrate with entrapped adsorbate layer.

and which Sears has more recently considered,[306] illustrates the most likely effect of impurity on crystal growth by a nucleation mechanism.*

In case 6 the surface contamination would affect the nucleation equation (D-37) by introducing a term α_{c_5} into the frequency factor, equation (D-34), and by lowering the value of the surface free energy to σ'. Hence instead of equation (D-37) one would obtain for the critical supersaturation

$$(p/p_e)''_{\text{crit}} = \exp\,(\pi h \Omega \sigma'^2/65\alpha_{c_5} k^2 T^2). \qquad \text{(D-68)}$$

The effect on the frequency factor would tend to cause $(p/p_e)''_{\text{crit}}$ to exceed $(p/p_e)_{\text{crit}}$ while the effect† on σ' would do the opposite. Also if $(p/p_e) > (p/p_e)''_{\text{crit}}$, α_{c_5} could still multiply the ideal growth equation. Hence the behavior illustrated in Fig. 29 for entropy constraints would be expected also for surface contamination constraints. However, $(p/p_e)''_{\text{crit}}$ could be greater or less than $(p/p_e)_{\text{crit}}$, depending on the relative importance of σ' and α_{c_5}. It is emphasized that equation (D-68) holds only for case 6 in Fig. 31. Other cases would have differing kinetics.

Impurities adsorption may also affect the topography of a growing crystal.

* Cabrera[307] has erroneously stated that Sears[306] and Ives and Hirth[299] predict that a lowering of σ' would decrease the nucleation rate. As these authors[306,299] demonstrated and as shown in equations (D-34) and (D-68) a lowering of σ' clearly would increase J and decrease $(p/p_e)''_{\text{crit}}$.

† Sears[306] has extended equation (D-68) to cover the case where complete monostep adsorption occurs by introducing the equivalent of the Gibbs adsorption isotherm.

Let us refer once again to Frank's theorem[274] in Section (D-3). If an impurity is adsorbing onto a crystal in a time-dependent manner and if the impurity slows down the ledge, as would be the usual case, then $\partial^2|J_c|/\partial(1/\lambda)^2$ is positive. That is, the greater the distance to the ledge leading a given ledge the slower the latter ledge will travel. This leads to an instability in a train of equally spaced ledges moving across a surface in condensation. If by a fluctuation one ledge spacing is increased, the following ledge will decelerate and the ledges following will pile up against the decelerating ledge, forming a macroscopic step. Thus when such adsorption takes place, condensation proceeds by the motion of macroscopic steps which are separated by essentially perfect low-index planes.

In such a circumstance, the condensation coefficient for the flat regions between steps will differ from that for the "risers" or high-index planes forming the steps of height h. Thus the exact condensation coefficient in such a circumstance will depend on the ratio of h to the width of the flat regions and on the values of α_{v_i} for the step and the riser, respectively.

If an impurity or some other constraint leads to macroscopic step formation, it is clear that equation (D-58) will still describe the diffusion kinetics. However, as pointed out by Chernov,[308] the step velocity must be inversely proportional to step height. Hence for macroscopic steps equation (D-65) is replaced by*

$$v = -(2D_s\Omega/h_j)(\partial n_s/\partial x)_{x=0} \qquad \text{(D-69)}$$

where $j = 1, 2, \ldots$ is the integral number of molecular step units in a macroscopic step. The kinetics will otherwise remain the same as for monomolecular ledges.

2. *Theory of Crystal Evaporation*

The above treatment for formation of macroscopic steps also holds for evaporation, and again the condition for macroscopic step stability is

$$\partial^2|J_v|/\partial(1/\lambda)^2 > 0.$$

With regard to microscopic kinetics, equation (D-58) will still describe the evaporation kinetics when adsorption occurs. However adsorption at a ledge would reduce n_s' by increasing $\Delta H^{\star}_{\text{k-1}}$ and $\Delta H^{\star}_{\text{1-ad}}$ to $\Delta H^{\star\prime}_{\text{kl}}$ and $\Delta H^{\star\prime}_{\text{1-ad}}$, respectively. In such a case, instead of equation (D-63), one would obtain†

$$\alpha_{v_8} = \alpha_\lambda f(\text{imp}) f(\delta_{\text{1-ad}})/[(\alpha_\lambda + f(\delta_{\text{1-ad}})] \qquad \text{(D-70)}$$

where

$$f(\text{imp}) = \exp\,[-(\Delta H^{\star\prime}_{\text{k-1}} + \Delta H^{\star\prime}_{\text{1-ad}} - \Delta H_{\text{kl}} - \Delta H_{\text{1-ad}})/kT] \qquad \text{(D-71)}$$

* For very large step heights a term involving the direct impingement on the riser, i.e. $ja\alpha_{c_i}(p - p_e)/(2\pi mkT)^{\frac{1}{2}}$, would have to be added to equation (D-69). Usually $\bar{X} > ja$ and this term can be neglected.

† As indicated in Section E, there is an equivalent $\alpha_{c_8} = \alpha_{v_8}$.

Finally, as noted by Stranski,[271,305] if the impurity chemisorbs it could set up a surface dipole which would have the effect of "protecting" crystal edges, i.e. preventing the edges from acting as ledge sources at low undersaturations. If the edges are protected, two-dimensional nucleation, as predicted by equation (D-52), may occur on perfect crystal faces. As discussed above for crystal growth, the principal effect of impurity adsorption should be to lower ε in equation (D-52) and hence to enhance the nucleation process.

3. *Experiment*

Numerous experiments have been carried out in which macroscopic steps have formed in growth or evaporation and in which α_{c_i} or $\alpha_{v_i} < 1$, due apparently to impurity effects. However, in most of these cases it is likely that the crystals were imperfect. Hence further discussion of this point is deferred to the next section where it will be shown that the above treatment is adaptable to the case of imperfect crystals.

In one pertinent experiment on a presumably perfect surface Dittmar and Neumann[309–312] observed potassium whisker growth at 56°C in a vacuum which is estimated to be $\sim 10^{-7}$ mm Hg. They observed whisker growth at $1.05 < (p/p_e) \precsim 1.7$, while the value calculated from equation (D-37) is $(p/p_e)_{\text{crit}} \simeq 2$. Their finding is in agreement with expectation for a clean perfect surface. However, they observed evaporation only from the tip for $0.48 < (p/p_e) < 1.0$, while evaporation ledges also originated at the prism edges for $(p/p_e) < 0.48$. This is in disagreement with Sears'[293,295] findings and with theoretical expectation.[262] The disagreement has been interpreted by several authors[313,314] as being due to adsorption of an impurity on the potassium prism faces which set up a surface dipole and thus protected the crystal edges.[271] This point would seem to merit further investigation.

Several other significant experiments have been performed on the effect of "poisons", or adsorbed impurities, on crystal growth from dilute aqueous solution. Although this process is not a subject of the present book, it is analogous to crystal growth from the vapor. In the growth of lithium fluoride platelets from dilute aqueous solution, Sears[306] has shown that the rate of increase in thickness of the platelets is increased by addition of 2 parts per million of ferric fluoride to the solution. Sears attributes this effect to a lowering of σ to σ', equation (D-68), and hence to a lowering of $(p/p_e)_{\text{crit}}$ to $(p/p_e)''_{\text{crit}}$. Sears[315] has also shown that ferric fluoride both enhances the rate of two-dimensional nucleation on low-index surfaces and slows the lateral propagation rate of the growing layers. This finding supports the view that case 6 in Fig. 31 described the ferric fluoride impurity effect.

Similarly, Sears[316] has studied the nucleation and growth of potassium chloride crystals from dilute aqueous solutions containing zero, 2 ppm and 10 ppm of lead chloride "poison". He found that in the absence of lead chloride, single platelets grew laterally but not in thickness, presumably because the basal surfaces were perfect. However, with the addition of 2 ppm of lead chloride, single platelets thickened at a finite rate, even though the supersaturation was the same for the pure solution. Again this is attributed to a lowering of σ to σ' in equation (D-68).

On the other hand, Michaels and Colville[317] found that adsorbate decreased the growth rate of adipic acid from aqueous solution. Their results appear to be consistent with two-dimensional nucleation theory* because a plot of the log of the growth rate R vs. the reciprocal of the log of the supersaturation ratio gives a straight line in agreement with equation (D-34). Thus their results may have been due to the effect of α_{c_5} in equation (D-68). However, it seems more likely that case 4 or 5 obtained in their experiment, rather than case 6 of Fig. 31. Further evidence for growth by two-dimensional nucleation and for monolayer adsorption of impurity, as required by case 4 or 5, is provided by their observation on the effect of surface active adsorbate. An anionic adsorbate selectively restricted growth of (110) and (010) faces, leading to formation of whiskers or rods with (001) axes. A cationic adsorbate had the opposite effect, leading to formation of platelets.

In summary, impurity adsorption should lead to formation of macroscopic ledges in crystal growth or evaporation and should lower the rate of crystal evaporation. However, it may either raise or lower the rate of crystal growth. None of these predictions has been experimentally demonstrated for growth of a perfect crystal from the vapor. Nevertheless some evidence obtained in growth from solution supports the theory.

* More recent work by Michaels and Tausch[318] on the same system but at lower supersaturation indicates that there are two linear ranges of different slope in the plot of $\log R$ vs. reciprocal log supersaturation, casting some doubt on the above discussion. However, this discussion may still be valid if the growth[318] at lower supersaturation represents a double-dislocation-source mechanism (see next section), as suggested by Michaels and Tausch,[318] while the growth at higher supersaturation is controlled by two-dimensional nucleation. In this respect it is puzzling that the growth rate does not approach the equivalent of the ideal equation (A-2) at the highest supersaturations. Continuing work by Michaels[319] should clarify these issues.

E. GROWTH AND EVAPORATION OF IMPERFECT CRYSTALS

1. Introduction

In the previous section it was shown that in most instances ideal crystals grow or evaporate by a mechanism of (for growth) adsorption, surface diffusion and accretion to monomolecular ledges on crystal surfaces. The kinetics of ledge motion were developed for ideal crystals.

Now imperfect crystals follow the same kinetic laws and hence evaporate and grow in accordance with the equations developed in the previous section. The only difference in kinetic behavior of imperfect crystals is due to the fact that the spacing between ledge sources, such as cracks, grain boundaries, crystal edges and dislocations, is finite and small. Therefore deviations occur from the asymptotic steady-state behavior characteristic of large low-index surfaces of perfect crystals. Thus in this section only the effects of imperfections and impurities on the *spacing* of monomolecular ledges will be introduced as new theoretical material; all matters relating to molecular kinetics were treated in the previous section. In sequence we will consider growth kinetics of imperfect crystals, the effect of impurities on growth kinetics and on crystal topography, evaporation of imperfect crystals and finally the effect of impurities on crystal evaporation.

Reviews on this subject have been written by Burton *et al.*,[(260)] Knacke and Stranski,[(251)] Hirth and Pound,[(41)] Chernov,[(320)] Dekeyser and Amelinckx,[(321)] Verma[(322)] and Courtney.[(323)]

2. Crystal Growth of Imperfect Crystals

(a) *Theory*

1. *Clean Imperfect Crystals*

(a) *Kinetics.* The induction that spiral* dislocations intersecting free surfaces are responsible for crystal growth from the vapor at low supersaturations was made by Frank[(324)] in 1949. When a large surface containing

* A spiral dislocation is defined as a dislocation intersecting a free surface and having a component of its Burgers vector normal to the free surface. Only these dislocations are important in crystal growth. Dislocations with Burgers vectors lying in the surface will not have a ledge associated with them. Hence a surface intersected by these latter dislocations would have to grow by a two-dimensional nucleation process as discussed in Section

one spiral dislocation is exposed to a supersaturation, the ledge, which terminates at the dislocation and at an edge of the surface, will advance by accretion of diffusing adatoms. However, the ledge will be constrained to terminate at the dislocation, so that the ledge will wind up into a spiral as shown in Fig. 32. The limit of the curvature of the spiral will be determined

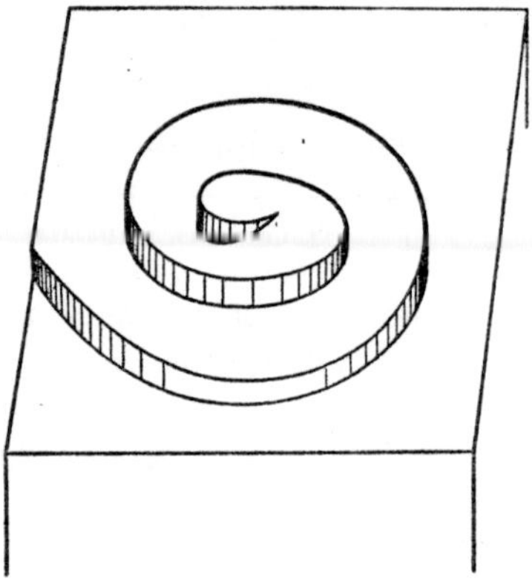

FIG. 32. Spiral-dislocation ledge on a growing crystal.

by the Gibbs–Thomson relation, i.e. the curvature of a ledge in equilibrium with the supersaturated vapor. Thus at the dislocation, the radius of curvature*

$$\rho_c = -\sigma/\Delta G_v. \tag{E-1}$$

With this boundary condition and assuming steady-state spiral rotation, Burton, Cabrera and Frank[260] showed that the spacing between spiral ledges emanating from a dislocation is approximately $\lambda_1 = 4\pi r^\star$, and Cabrera and Levine[325] later corrected this value to†

$$\lambda_1 = 19\rho_c = -19\sigma/\Delta G_v. \tag{E-2}$$

D. There is a positive contribution to the free energy of formation of a disc-shaped nucleus at the latter type of dislocation because of the elastic-strain requirement to accommodate the dislocation. This type of dislocation is a *less* favorable growth site than a surface region free of dislocations. General usage in the growth literature is to call a spiral dislocation a "screw" dislocation. However, a screw dislocation is defined as a dislocation with its line and Burgers vector parallel.[266] This does not correspond to the above definition of a spiral dislocation, so we shall use the term spiral dislocation to describe the dislocations which promote spiral crystal growth.

* Of course this equation implies isotropic edge energy for the ledge. If the line energy were markedly anisotropic, ρ_c would be replaced by a critical length corresponding to the smallest straight length of low-index edge adjacent to the dislocation.

† Strictly speaking, this condition applies only for spiral dislocations in which the Burgers vector is small and the surface energy is large, i.e. when $\mu b^2/8\pi^2\sigma \gtrsim a$, where μ is the shear modulus and b is the magnitude of the Burgers vector of the dislocation.[266]. This condition usually obtains in crystal growth. The only difference in growth when this condition does not obtain would be a perturbation in ledge spacing at a spiral source.

Thus for a case in which surface diffusion enters the kinetics* (cf. equation D-58), they obtained the relation

$$\alpha_\lambda = (\sqrt{2}\bar{X}/\lambda_1)\tanh(\lambda_1/\sqrt{2}\bar{X}). \tag{E-3}$$

For example, a clean metal crystal with a monatomic vapor phase would follow the growth law

$$\alpha_{c_9} = (\sqrt{2}\bar{X}/\lambda_1)\tanh(\lambda_1/\sqrt{2}\bar{X}) \tag{E-4}$$

and would predict the growth behavior depicted in Fig. 33. The growth rate should be ideal ($\alpha_c = 1$) if $\lambda_1 < \bar{X}$. On substituting into equation (E-2) it is seen that the growth rate should be ideal for

$$(p/p_e) > (p/p_e)_1 = \exp(19\sigma\Omega/\sqrt{2}\bar{X}kT). \tag{E-5}$$

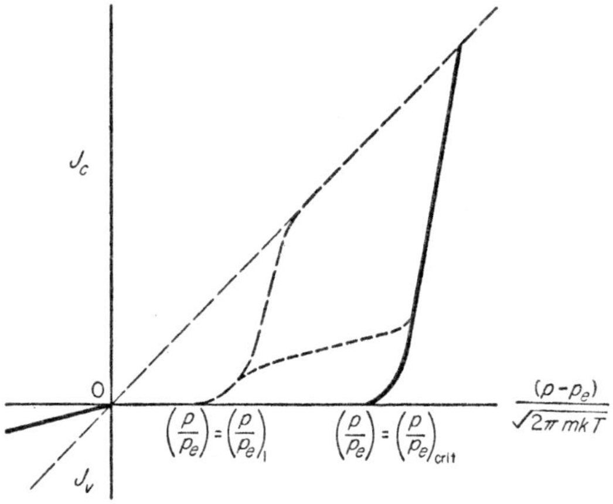

FIG. 33. Theoretical growth rate versus ideal growth rate for a large but finite crystal, bounded by low-index planes each containing one spiral dislocation: — — — as predicted by Burton *et al.*,(260) - - - - as predicted by Hirth and Pound.(41,326) The growth rate ——— for a perfect crystal is also shown.

However, Hirth and Pound(41,326) showed that when $\lambda_1 < \lambda_0$, as given by equation (D-67), ledges tend to accelerate at a decreasing rate and approach an asymptotic state. As discussed in Section D-3, for the asymptotic state of the clean metal

$$\alpha_{c_6} = \alpha_\lambda \cong \sqrt{2}\bar{X}/\lambda_0 = \tfrac{1}{3}. \tag{E-6}$$

On the other hand when $\lambda_1 > \lambda_0$, i.e. when $(p/p_e) \gtrsim (p/p_e)_1$, equation (E-3), as given by Burton *et al.*, should hold. Also, at high supersaturations where $(p/p_e) \gtrless (p/p_e)_{crit}$ (see equation D-37), the ideal growth rate should obtain either because of rapid nucleation or onset of diffuseness of the interface(261)

* In any case where surface diffusion enters the kinetics, which includes almost all the possibilities described by equation (D-58), similar considerations would lead to a change in kinetic behavior at the supersaturation where $\lambda_1 = \lambda_0$.

as discussed in Section D. This latter postulated behavior is also shown in Fig. 33 for a simple metal crystal. Of course where entropy effects are present (see equation D-58), α_c will not become unity at $(p/p_e)_{crit}$ but will approach the value determined by the entropy constraint.

Figure 33 refers to a large low-index surface containing one spiral dislocation. If there were an equal number of dislocations of opposite sign intersecting the surface and the distance between pairs of these were d, growth would not occur unless the supersaturation were sufficiently high that $2r^\star < d$,[260] as illustrated in Fig. 34. Also, if $\lambda_1 > \lambda_0$, steady-state behaviour

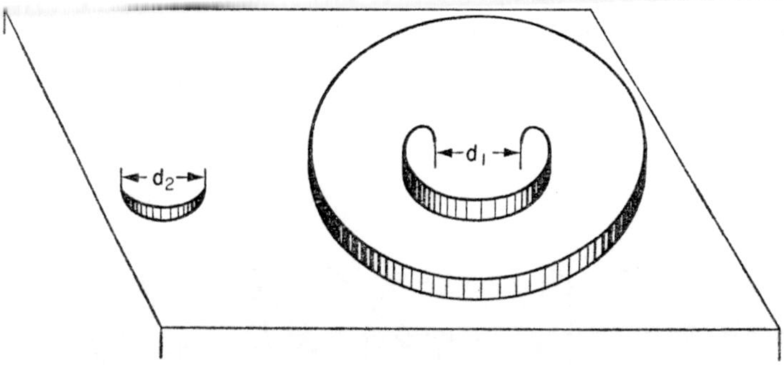

FIG. 34. Crystal surface containing spiral dislocations where $d_1 > 2r^\star > d_2$.

will obtain and α_c will approach 0 for $\lambda_1 \gg \lambda_0$. On the other hand if $\lambda_1 \ll \lambda_0$, ledges will be more closely spaced near the dislocation source, varying from λ_1 near the source to λ_0 at about $5\lambda_0$ distant.[326] In the latter case α_c

TABLE 10

FRACTION OF AREA AFFECTED BY CAPILLARITY A_a AND TOTAL FRACTION OF PERTURBED AREA A_0 BASED ON A SPIRAL-DISLOCATION CONCENTRATION OF 10^5 PER CM^2

Metal	T°K	σ (ergs/cm^2)	A_a ($\times 10^4$)	A_0
Na	250	300[327]	3.46	1
	350		1.92	0.58
Mo	1400	2530[327]	1.31	1
	2700		0.35	0.002
Fe	1200	2040[327]	0.65	0.62
	1800		0.30	0.002
Cd	350	410[327]	1.04	0.009
	500		0.51	0.0002
Cu	800	1650[328]	0.97	0.51
	1300		0.37	0.0008
Ag	800	1130[328]	0.96	0.03
	1200		0.43	0.0003

will vary from the value given by equation (E-3) near the source (~ unity when $\lambda_1 < \lambda_0$) to the value given by equation (E-6) far from the source. Thus if the dislocation density is low the total area of perturbed region around dislocations will be small and $\alpha_{c_6} \simeq \frac{1}{3}$. If the dislocation density is high, the perturbed region will be large and $1 > \alpha_c \gtrsim 0.3$. The results of a calculation of the perturbed area for various metals is presented in Table 10.[326]

(b) *Topography.* The presence of spiral dislocations in crystal growth can have several effects on growth topography. First of all as noted above, if $\lambda_1 < \lambda_0$ the spacing of ledges will increase with distance from a spiral center, resulting in the formation of a growth cone (Fig. 35). As suggested by

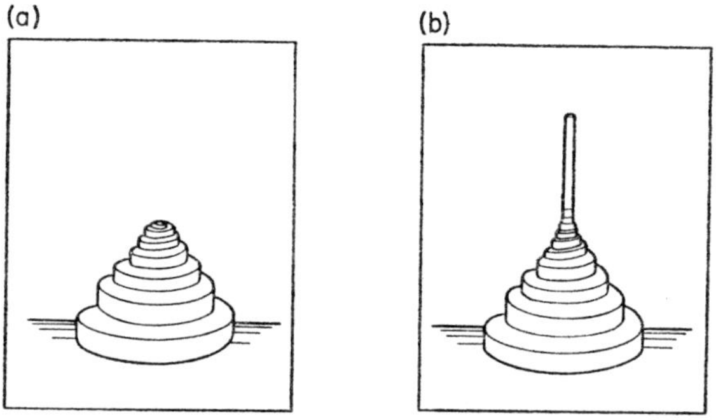

FIG. 35. Whisker (b) emanating from a growth cone (a).

Brenner and Sears,[329] if the crystal is growing in the presence of an inert gas such that a diffusion gradient exists in the vapor, the tip of the cone will penetrate a region of higher supersaturation than the base and should hence grow faster. As the tip grows faster the diffusion field about the tip should become nearly spherical. Eventually the tip of the spiral would become a whisker growing with perfect sides, but with a dislocation spiral at the tip as shown in Fig. 32.

Similarly, if by some growth accident the lateral motion of spiral ledges were restricted at some distance from a source, the ledges would accumulate at the constriction and once again a whisker would form. In this case the radius would be equal to the distance between the dislocation ledge source and the constriction.[282]

Finally, it has been shown by Frank[330] and Forty[331] that spiral-dislocation growth may affect crystal structure. The Burgers vector of the dislocation[266] is supposed to be determined either by the substrate on which the growing crystal nucleated or by a growth accident. The component of this Burgers vector normal to the surface will determine the effective step height of the spiral dislocation, while the Burgers vector will determine the positioning of the molecules added in one turn of the spiral. For example in the case of a metal, a f.c.c. lattice can be generated by stacking close-packed layers of height equal to $(2/3)^{\frac{1}{2}}$ times the distance d of closest approach of the atoms in the sequence ABCABCABC.[266] On the other hand, a hexagonal close-packed lattice can be generated by stacking layers of the same height in the sequence ABABAB. A stacking fault in the f.c.c. lattice would be represented by ABCABC$\downarrow$BCABCABC, and the h.c.p. lattice by ABABA$\downarrow$CBCBCB. Now if a h.c.p. crystal is growing normal to the basal plane by the spiral-dislocation mechanism and the Burgers vector component normal to the surface of a perfect dislocation is $(2n)(2/3)^{\frac{1}{2}}\ d$, where $n = 0,1,2,\ldots$, the crystal will grow in the correct stacking order ABABAB. However, if the vector component is $(2n + 1)(2/3)^{\frac{1}{2}}\ d$, a plane of stacking fault will be generated for each spiral revolution. For example if the vector component is $(2/3)^{\frac{1}{2}}\ d$, the stacking will be ABCABCABC, which means that every plane generated for a normally hexagonal crystal would be faulted. In the case of metallic crystals, the instability of large dislocations[332] suggests that only values of $n = 0$ to 2 are likely possibilities for the above vector components. For other materials larger spiral-dislocation vector components are possible.

As pointed out by Cabrera,[333] the generation of a stacking fault by one of a set of monomolecular ledges originating at a spiral dislocation of large Burgers vector will reduce the driving force for advance of that ledge and hence retard its forward motion. The other monomolecular ledges will pile up behind the slow ledge so that a macroscopic step will form. Thus the generation of stacking fault is an alternative to impurity adsorption in the formation of a macroscopic step.* This effect is not likely to occur in metals because of the instability of dislocations of large Burgers vector.

2. *Impurity Effects*

(a) *Kinetics.* The effect of adsorbate on the motion of monomolecular steps is as discussed in Section D-4 for evaporation. The principal effect should be to reduce the kinetics at ledges by reducing n_s' because adsorption should increase $\Delta H^{\star\prime}_{\mathrm{k-l}}$ and $\Delta H^{\star\prime}_{\mathrm{l-ad}}$ to values much larger than $\Delta H_{\mathrm{k-l}}$ and $\Delta H_{\mathrm{l-ad}}$, respectively. Accordingly from equation (D-58)

$$J_c \cong (\alpha_\lambda \nu/\delta_{\mathrm{k-l}}\delta_{\mathrm{l-ad}}a^2)\exp(\Delta H^{\star\prime}_{\mathrm{k-l}}/kT). \qquad \text{(E-7)}$$

A transition from $\alpha_\lambda = 1$ to $\alpha_\lambda < 1$ will occur (cf. equation E-5) at

$$(p/p_e) < (p/p_e)_1' = \exp(19\sigma'\Omega/\sqrt{2}\bar{X}kT), \qquad \text{(E-8)}$$

in which the dependence of σ' on adsorption has been introduced. Also when $(p/p_e) > (p/p_e)''_{\mathrm{crit}}$ (see equation D-68), surface contamination will still lower the net flux of condensation. Hence when impurity adsorption occurs and monomolecular-ledge kinetics obtain, the kinetics of Fig. 36 should replace those of Fig. 33.

* Chernov[320] and W. W. Mullins (private communication) have pointed out that if two monatomic ledges have a negative energy of interaction this may lead to their coalescence to form a macroscopic step.

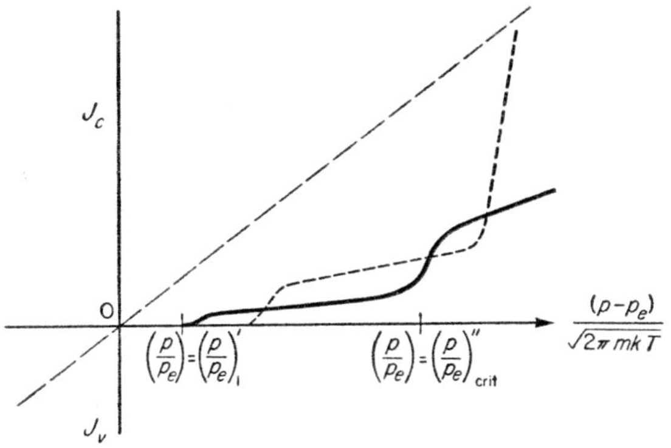

FIG. 36. Growth rate versus ideal growth rate for a large but finite crystal, bounded by low-index planes, each containing one spiral-dislocation: ——— in the presence of impurity adsorbate; – – – – for clean crystal.

(b) *Topography.* If adsorption of impurities at ledges takes place in a time-dependent manner, stable macroscopic steps may arise as proposed by Frank[258] and noted in Section D-4. Lang[334] has shown that if selective bunching of monomolecular ledges occurs at one side of a circularly symmetric source, a macroscopic spiral will form regardless of whether the source is a spiral dislocation. However, it is felt that most macroscopic spirals should originate from spiral-dislocation sources. The detailed kinetics of crystal growth are as outlined in Section D-4. There it was shown that when macroscopic steps are present one must consider the effective condensation coefficients of both the flat regions and the "risers" of the steps.* However, the ledge spacing evidently can no longer be predicted either from diffusion theory[41] or from the effect of capillarity at the spiral source.[325] Hence the kinetics of crystal growth cannot be predicted in detail when adsorption has the above consequences. If topographical evidence for adsorption is present, this is an indication that simple kinetic laws are not likely to be followed in such crystal growth.

Referring to the previous section, it is also evident that time-dependent impurity adsorption could promote whisker formation. Near the dislocation source clean monomolecular ledges are being generated. These ledges slow

* Cabrera[335] has noted that in the extreme case, if the riser of a macroscopic step is a perfect crystal surface, two-dimensional nucleation could be required to advance the riser. In general the riser is a roughened surface, or itself may be intersected by dislocations, so that in such instances this effect does not obtain. Chernov and co-workers[320,336] have developed this idea extensively. They considered both the enhanced nucleation at the re-entrant angle at the base of the step[335,336] and the rate of supply of material to the step by surface diffusion. They concluded that macroscopic steps should break up into smaller ledges at some critical supersaturation. This break-up should be accompanied by an increase in growth rate, both because the ledge velocity should increase as step height decreases (equation D-69) and because the ledge spacing decreases. However, they did not consider the role of impurities, which appears to be critical in macroscopic-ledge formation.[258]

down and bunching may occur as impurity adsorption takes place. If the bunching takes place in a circularly symmetric fashion, a whisker will be generated in which the symmetric macroscopic-step risers become the perfect prismatic faces of the whisker.

Two mechanisms* have been suggested by which impurities could generate spiral dislocations in a growing crystal. Frank[338] postulated that if a crystal growing as a thin platelet absorbed impurities non-uniformly as it grew, the impurity gradient would set up shear stresses. These stresses would then cause the platelet to shear parallel to the crystalline axis of smallest dimension and hence to generate a line of spiral dislocations. Fisher *et al.*[339] proposed that two crystals may impinge upon one another with a small misorientation, giving rise to a line of spiral dislocations at their juncture. Alternatively, a macroscopic impurity particle may be incorporated into a growing crystal in such a manner that the crystal does not join perfectly after growing around the impurity particle. This could produce a spiral-dislocation step originating at the impurity. Newkirk and Sears[340] pointed out that if one or several such spiral dislocations are generated on only one set of faces of a small crystallite, a rod or whisker will form. On the other hand if such spiral dislocations are generated on two faces, effectively forming a small-angle twist boundary,[266] a platelet will form. Ehrlich[341] has noted that selective impurity adsorption could also lead to rod formation if the impurity selectivity restricted lateral-growth faces of a small crystallite.

Cabrera and Vermilyea[276] pointed out that certain adsorbed impurity molecules or molecular clusters may be immobile on a surface and that these may be effective in restraining motion of monomolecular ledges. Such a ledge would be stopped by impurities less than $2\rho_c$ (see equation E-1) apart. Now the density of impurity particles on a surface should be a function of the crystal growth rate (i.e. of the supersaturation) and of J_{imp}, the rate of impurity adsorption. Cabrera and Vermilyea, assuming a uniform distribution of impurity, showed that under the above conditions crystal growth should be stopped by impurity adsorption below a supersaturation

$$[(p - p_e)/p_e] = \text{constant} \times (J_{\text{imp}})^{\frac{1}{3}} \qquad \text{(E-9)}$$

3. *Whisker Growth*

The kinetics of whisker growth are treated separately because, as noted above, whiskers can be engendered by a variety of mechanisms. Once a whisker has formed it should grow by a mechanism of adsorption from the vapor, diffusion along the prism sides of the whisker to the tip and accretion

* Forty[331] has suggested that convection currents could cause shear stresses sufficient to buckle a platelet growing from solution. This mechanism is not likely, however, to occur in growth from the vapor. It also seems conceivable that a non-equilibrium supersaturation of vacancies could be grown into a crystal.[337] These might condense as a vacancy disc which would collapse to a dislocation loop. The loop could then slip to the surface of a growing crystal and become a spiral-dislocation source upon intersecting the surface.

of molecules into the spiral dislocation at the tip. Since $(\bar{X}/\tau_s) \gg \dot{l}$* the axial growth velocity of the whisker, the moving boundary condition on the surface can again be neglected. Thus the surface-diffusion differential equation is identical with equation (D-56). In this case the boundary conditions are

$$(\partial n_s/\partial x) = 0$$

at $x = 0$, $n_s = n'_s$ at $x = l$.† Analogously to the derivation of equation (D-57), one can solve equation (D-56) with the above boundary conditions and obtain $n_s(x)$ and $(\partial n_s/\partial x)_{x=l}$. The whisker growth rate is given by

$$\dot{l} = (C/A)\Omega D_s(\partial n_s/\partial x)_{x=l} + AJ_{\text{tip}} \qquad \text{(E-10)}$$

where C is the circumference of the whisker, J_{tip} is the net impingement flux on the whisker tip, and A is the area of the tip. In most cases A is small and the second term on the right of equation (E-10) is negligible. Accordingly, substituting for $(\partial n_s/\partial x)_{x=l}$ in equation (E-10) one obtains (cf. equation D-4)

$$\dot{l} = (C/A)\Omega(\bar{X}/\sqrt{2}) \tanh(\sqrt{2}l/\bar{X})[\delta_{\text{ad}}p/(2\pi mkT)^{\frac{1}{2}} - n'_s\nu \exp(-\Delta H^{\star}_{\text{des}}/kT)]. \qquad \text{(E-11)}$$

In this simple case of a clean crystalline surface with a monatomic vapor phase $\delta_{\text{ad}} = 1$ and it is likely that $n'_s = n_{s_e}$ and $\Delta H^{\star}_{\text{des}} = \Delta H_{\text{des}}$. Thus equation (E-11) reduces to

$$\dot{l} = (C/A)\Omega[(\bar{X}/\sqrt{2}) \tanh(\sqrt{2}l/\bar{X})](p - p_e)/(2\pi mkT)^{\frac{1}{2}} \qquad \text{(E-12)}$$

which is identical with the solution derived for the above conditions by Dittmar and Neumann.[312] In the limits of small and large l, respectively, equation (E-12) reduces to the relations

$$\dot{l} = (C/A)\Omega l(p - p_e)/(2\pi mkT)^{\frac{1}{2}} \qquad \text{(E-13)}$$

and

$$\dot{l} = (C/A)\Omega(\bar{X}/\sqrt{2})(p - p_e)/(2\pi mkT)^{\frac{1}{2}}. \qquad \text{(E-14)}$$

Equation (E-13) has been derived by Gomer[153] and (E-14) by Sears[282] independently of the analysis leading to (E-12).

The cessation of rapid axial whisker growth has been postulated to occur when: (i) the spiral dislocation(s) responsible for whisker growth slip[342] or climb[337] out of the whisker during the course of crystal growth; (ii) an impurity adsorbs at the spiral-dislocation ledge at the whisker tip;[276] or (iii) two-dimensional nucleation commences on the prism faces of the whisker.[153,282] The first possibility is likely only for whiskers of small radius, while the second is likely for any whisker size. The third possibility should

* τ_s is the mean residence time of an adatom.
† n_s is the adsorbed population in equilibrium with the vapor.
n'_s is the adsorbed population in equilibrium with a monatomic ledge.

occur only late in whisker growth (where $l > \bar{X}$). In any case slip or climb of the dislocation out of the whisker or impurity adsorption at the tip may also occur when the third possibility obtains.

Gomer[153] suggested that the surface diffusion coefficient could be determined by noting the length at which exponential whisker growth ceased. However, as shown by Dittmar and Neumann[312] only $\bar{X}$, (equation D-38), which is insufficient to determine D_s, can be determined from such a measurement, and hence we regard with reservation the surface-diffusion coefficients determined in this manner. Also, in another procedure the terminal length is taken as the length at which $l = \bar{X}$ and thus where rapid whisker growth ceases because of two-dimensional nucleation on the side faces.[153,282] This terminal length is then used to calculate the surface-diffusion coefficient as noted above. We have reservations about this method also, both because of the considerations mentioned above and because of the other possible mechanisms[276,337,342] of whisker-growth cessation.

For the case of platelets the kinetics of growth are similar to those of whisker growth except that the crystal is growing in two dimensions. Sears[16] has shown that the growth equation for platelets, analogous to equation (E-14), is

$$\dot{y} = (2/b)\Omega(\bar{X}/\sqrt{2})(p - p_e)/(2\pi mkT)^{\frac{1}{2}} \tag{E-15}$$

where b is the platelet thickness and y is the distance from the platelet center to the midpoint of a growing edge. For more complex crystals an equivalent modification of equation (E-11) applies.

As shall be discussed in the experimental section, there is some controversy as to whether whiskers and platelets indeed contain spiral dislocations. A conceivable alternative is that two-dimensional nucleation is easier on one set of low-index crystallographic planes than on others, either because of a lower surface free energy for the latter planes or because of impurity adsorption. In either case equations (E-11) and (E-15) will still describe the kinetics as long as nucleation on the growing faces is rapid enough to maintain the equilibrium concentration of adsorbed molecules at the edges where the growing faces intersect the dormant ledges.

A special case, suggested by Forty,[343] of perfect crystal growth could occur if, say, the lateral faces of a platelet broke up into a hill-and-valley faceted surface in order to minimize surface free energy in the manner proposed by Herring.[101] If this occurred nucleation would be enhanced at the re-entrant corners formed by the facets and one would expect growth of such an interface at a lower supersaturation than that required for growth of a perfect flat surface.

Another possibility which does not appear to have received attention is that a tiny crystallite may form such that a coherent twin plane is incorporated into the crystal by a mechanism such as that leading to equation (D-41). The lateral faces intersected by the twin could then develop a re-entrant angle where two low-index surfaces intersect at the line of emergence of the twin interface. Such a re-entrant edge could then promote nucleation on the lateral faces as discussed above.

(b) *Experiment*

1. *Kinetics*

(a) *Low supersaturations.* The discussion of the kinetics is divided into two sections: (i) low supersaturations where $(p/p_e) < (p/p_e)_1$ (see Fig. 33) and screw-dislocation spiral ledges control growth kinetics, and (ii) high supersaturations where $(p/p_e) > (p/p_e)_{crit}$ and growth should be dominated by two-dimensional nucleation.

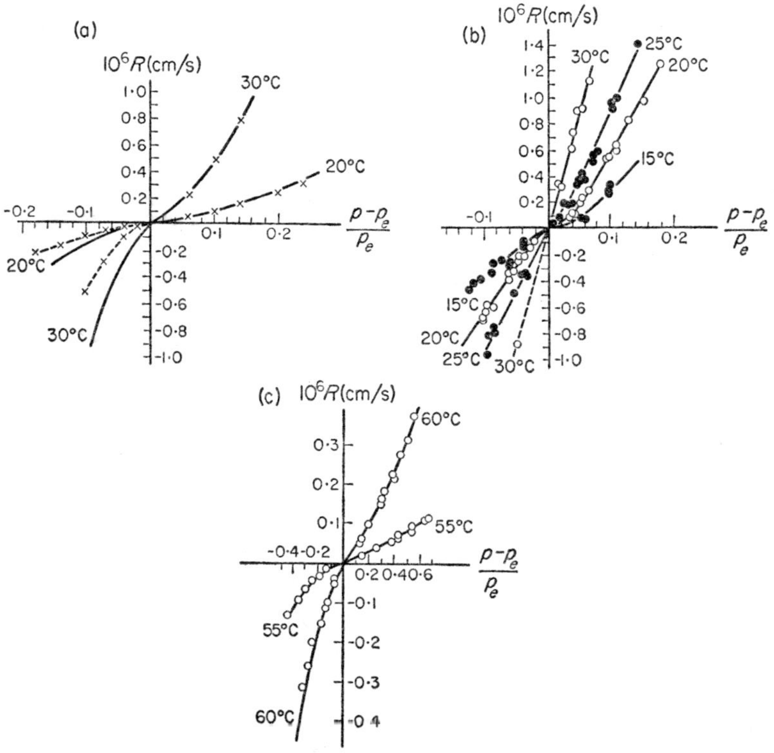

FIG. 37. Growth and evaporation rates R of (a) benzophenone (110) face, (b) benzophenone polycrystal, and (c) rhombic sulfur (111) face. x—x—x represents rate calculated from equation (D-58).

Kitchener and Strickland-Constable[344] have studied the growth of macroscopic crystal faces* of benzophenone and rhombic sulfur at low supersaturations with fluxes of $\sim 10^{16}$ molecules/cm²/sec. Some of their results are presented in Fig. 37. They compared their results with the equation

* The large single crystals which they used probably had screw dislocations on all faces.

of Burton *et al.*,[260] which corresponds to equations (D-63) and (E-4) or (E-6) with $f(\delta_{1-\text{ad}}) > \alpha_\lambda$ and to the curve in Fig. 33. As shown in Fig. 37 they found parabolic growth behavior, in agreement with the theoretical predictions of either equation (E-4) or (E-6), when $(p/p_e) < (p/p_e)_1$. In the one case which they treated quantitatively, case (a) in the figure, they obtained excellent agreement with experiment upon fitting the experimental curve by equation (D-63) with $\delta_{\text{ad}} = 0.11$,* $(p/p_e)_1 = 1.1$ at 20°C and $\delta_{\text{ad}} = 0.135$, $(p/p_e)_1 = 1.1$ at 30°C. However, according to equation (D-63) a plot of $J_c(\mathrm{d}J_c/\mathrm{d}S)$ versus S, where $S = (p - p_e)/p_e$ is the supersaturation, should approach a straight line of 45° slope at large S. In contrast, the plot appears to diverge continuously at high S. Also the value of $(p/p_e)_1 = 1.1$ appears to be too small, because $\mathrm{d}^2R/\mathrm{d}S^2$ should become negative at $(p/p_e)_1$ according to Fig. 33 while from Fig. 37 $\mathrm{d}^2R/\mathrm{d}S^2$ is still positive at $(p/p_e) = 1.3$. Further it is just in the region of $(p/p_e) \gtrless (p/p_e)_1$ where differing kinetic behavior is predicted by equations (E-4) and (E-6) as shown in Fig. 33. Accordingly it would be quite interesting to extend the foregoing work to somewhat higher supersaturations. It is noted that the experiment was carried out in a vacuum of $\sim 5 \times 10^{-6}$ mm Hg and thus time-dependent contamination by residual gas was possible. Considering all of the above points, however, the agreement between theory and experiment is encouraging.

Bradley and Drury[345] studied the growth of edges of hexagonal platelets of iodine at 0°C and 25°C in a vacuum of $\sim 10^{-5}$ mm Hg. Analyzing their data as in the above work, they found agreement with equations (D-63) and (E-4) or (E-6) for $(p/p_e)_1 = 1.135$, $\delta_{\text{ad}} = 0.29$ at 25°C and $(p/p_e)_1 = 1.1$, $\delta_{\text{ad}} = 0.29$ at 0°C. In this same range of measurements, $1 < p/p_e < 1.07$ at fluxes of $\sim 10^{18}$ molecules/cm²/sec, Volmer and Schültze[346] studied the crystal growth of iodine and obtained the results given in Fig. 38. This latter work would appear to accord with equations (D-36) and (E-6) with $(p/p_e)_1 \simeq 1.04$. However, the scatter in this work is greater than in that of Bradley and Drury. Nevertheless, a reservation about Bradley and Drury's work is noted: the growth rates may not correspond to growth of a plane surface because they measured the edge growth of small platelets. In this connection the reader is referred to the arguments leading to equation (E-15). Also impurity effects again may have been present. As in the benzophenone work, it would be interesting to extend the iodine measurements to higher supersaturations.

Bradley and Drury[345] also studied cubic carbon tetrabromide growing as pyramids bounded by (100) faces. Dislocations were probably present on all faces and the data for $1 < (p/p_e) < 1.04$, fluxes of $\sim 10^{18}$ molecules/cm²/sec and $T \sim 50$°C fit the above type of correlation with $\delta_{\text{ad}} = 0.0075$ and $(p/p_e)_1 = 1.1$. However, in the case of edge growth of platelets of monoclinic carbon tetrabromide they found linear growth with $\delta_{\text{ad}} \sim 0.04$ and $(p/p_e)_1 < 1.0004$ for $1.0004 < (p/p_e) < 1.03$ and $T = 40$°C. But in view of the fact

* δ_{ad} is the "free-angle" term in adsorption from the vapor.

that they were measuring edge growth of platelets, this result could be explained by equation (E-15) with $(\sqrt{2}\bar{X}/b)\delta_{ad} = 0.04$. Equation (E-15) would account for the linearity of the growth rate and for the apparent value of δ_{ad}, which was surprisingly higher than for the cubic form.

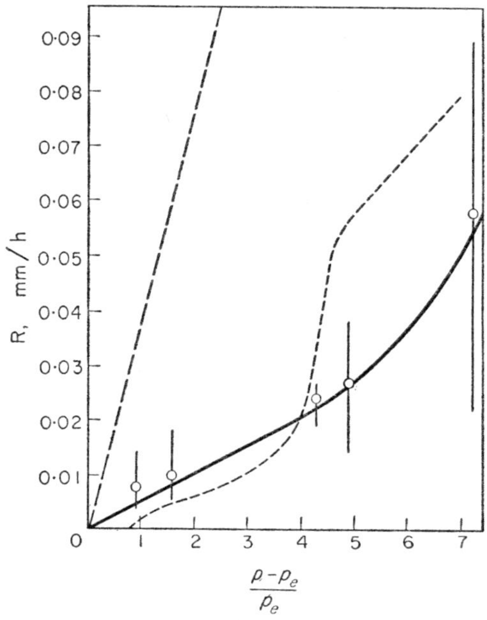

FIG. 38. Growth rate R versus supersaturation S for iodine; — — — ideal growth rate from equation (A-2), - - - - Volmer and Schültze,[346] ——— Bradley and Drury.[345]

In all of the above cases, if impurity adsorption were important equation (D-70) would apply instead of equation (D-63) and the factor f_{imp} would multiply δ_{ad} in each case. Particularly for carbon tetrabromide, which is a spherically symmetrical molecule and hence should not exhibit a large δ effect, the value of δ_{ad} seems quite low. Thus impurity adsorption may be a factor in the above work.

Volmer and Shultze[346] also studied the growth of P_4 at $T = 0°C$, fluxes of $\sim 10^{18}$ molecules/cm²/sec and $1.001 < (p/p_e) < 1.07$. As discussed by Burton *et al.*,[260] they found linear growth down to $(p/p_e) = 1.001$ and the data were fitted by $\delta_{ad} = 0.1$. Burton *et al.*[260] suggested that this indicates that $(p/p_e)_1 < 1.001$ for P_4. The result could also be explained, as in the preceding case, by equation (E-15).

McNutt and Mehl[347] studied the growth of cadmium at $1.5 \gtrsim (p/p_e) \gtrsim 5$ in a vacuum of $\sim 10^{-5}$ mm Hg. The fluxes were $\sim 10^{19}$ atoms/cm²/sec, substrate temperatures were 274 to 317°C and $(p/p_e)_{crit}$ was about 60. They observed interferometrically that cadmium grew by the lateral propagation on the basal plane of macroscopic steps. These steps may have been formed by time-dependent adsorption of some residual-gas impurity in their vacuum as discussed in Section D, although stacking faults[333] cannot be excluded as a possible cause of the macroscopic steps. They measured the lateral velocity of the macroscopic steps and found that it decreased with increasing step height in a manner consistent with equation (D-69). Thus they provided confirmation of the surface-diffusion mechanism of crystal growth upon which equation (D-69) is based.

Lemmlein *et al.*[348] performed similar experiments on the growth of naphthalene, diphenyl and *p*-toluidine at low, but not precisely known, supersaturations. Their vacuum was not cited but was probably no better than 10^{-5} mm Hg. They also found agreement with equation (D-69) for $j < 3$. However, for $j < 3$ the experimental values fell to about half of that predicted by equation (D-69), suggesting that perhaps Cabrera's[333] mechanism for hindrance of small steps by stacking faults might have been operative.

Parker and Kushner[349] measured the average growth rate of low-index facets of zinc crystallites at 390°C in a vacuum no better than about 10^{-4} mm Hg and at low supersaturations corresponding to $p/p_e = 1.009$ to 1.08. They found a linear dependence on supersaturations and $\alpha_c \simeq 0.1$. In view of the work of Rapp *et al.*[359] (discussed in the next section) who found $\alpha_c = 1$ for zinc at high supersaturation in a vacuum of 10^{-9} mm Hg and $T = 70$°C, Parker and Kushner's result can tentatively be explained by equations (D-63) and (E-6) and Fig. 33, or by equations (D-70) and (E-6) and Fig. 36. Again it would be of interest to extend this work to higher supersaturations so that the region of transition from $\alpha_c \simeq 0.1$ to 1 could be observed.

Chernov and Dukova[350] measured the crystal growth rate of β-methyl naphthalene and paratoluidine from the vapor at low supersaturation. Their vacuum was not specified but was probably no better than 10^{-5} mm Hg. They actually observed the motion of dislocation spirals during crystal growth. They found agreement with the form of Fig. 33 and with equations (D-63) and (E-6) for spirals of monomolecular step height. Also of great significance, they observed that spiral growth ceased and general flat-interface growth commenced at just the supersaturation corresponding to $(p/p_e)_{crit}$ in Fig. 33, *exactly* in agreement with the prediction of the model. For multimolecular step heights they found that the break in the growth rate versus supersaturation curve occurred at lower supersaturations. This may indicate enhancement of two-dimensional nucleation at the re-entrant corner formed at the bottom of the large steps.

Hallett[351] observed the growth of ice by the lateral propagation of macroscopic steps. He also found that the step velocity was inversely proportional to height, in agreement with equation (D-69). Also, by observing the time dependence of the growth of a circular step of known height and comparing the result with the expected rate of growth by a surface-diffusion mechanism, he calculated $\bar{X}$ to be about 7 microns at -6°C (see equation D-38). However, this value may not represent the true self-diffusion distance on ice, because the presence of macroscopic steps in his work suggests the possibility of impurity adsorption.

In the related problem of growth of β-methylnaphthalene from dilute ethanol solution, Kozlovskii and Lemmlein[352] measured the velocity of macroscopic steps as a function of supersaturation and height. Again, as shown in Fig. 39, they found agreement with equation (D-69); the velocity varied as the reciprocal of the step height and the velocity of a step of a given step height was directly proportional to the supersaturation. This latter requirement is occasioned by the expansion of equation (D-69) (see equation D-66) at low supersaturations where $\lambda_1 > \bar{X}$ and $\tanh(\lambda_1/\sqrt{2}\bar{X}) = 1$. Also, in agreement with equation (E-2), they showed that for a spiral of $j = 3$ the spacing λ decreased with increasing supersaturations as shown in Fig. 39.

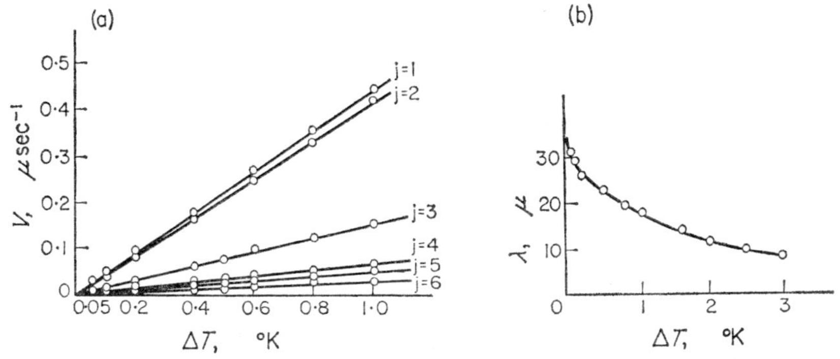

FIG. 39. (a) Velocity of propagation of macroscopic ledges of β-methylnaphthalene in growth from dilute ethyl alcohol solution as a function of supercooling ΔT (proportional to supersaturation); (b) spacing between successive ledges for a spiral of height $3a$ as a function of ΔT.

The finding that λ deviated from a hyperbola at large ΔT may indicate that the region where $\lambda_1 \leq \lambda_0$ was reached so that λ began to approach λ_0 asymptotically to approach the condition of equation (E-6).

Similarly, Kaischew *et al.*[353] showed that for ledges in a growth spiral the spacing λ decreased with increasing supersaturation in the electrolytic growth of silver at low overvoltages.

In summary, while agreement of theory with experiment is encouraging and certainly the surface-diffusion mechanism of crystal growth appears to have been verified, some points await clarification. All of the definitive quantitative experiments have been carried out in poor vacua and hence residual-gas adsorption could have affected growth kinetics. Also it would be of great interest to measure growth rates through the supersaturation range including $(p/p_e)_1$ and $(p/p_e)_{crit}$, because various theories predict differing kinetics at these supersaturation ratios. Further, one could calculate surface energies and surface diffusion coefficients from the above ratios and compare these parameters with independent measurement and/or theoretical prediction. Finally, the importance of viewing the growth morphology in order to discern whether growth is occurring uniformly over the entire crystal surface or only at platelet edges or whisker tips has been noted.

(b) *High supersaturations.* As discussed in Section C on heterogeneous nucleation from the vapor, in many cases after nucleation has occurred islands of growing crystallites will be present on the substrate. Frequently the crystallites will not impinge upon each other until the average thickness of the deposit is a few hundred molecular diameters. Thus the finding cited in Section C and elsewhere[184,354] that not all molecules striking a heterogeneous substrate stick when the total deposit is only a few molecular diameters thick does not prove that $\alpha_c < 1$ in such instances; the fraction of molecules that do not stick may have adsorbed and re-evaporated from a portion of the heterogeneous substrate where no growing nucleus was present. Similarly the finding that all molecules stick to the surface in such an instance strongly implies, but does not prove, that $\alpha_c = 1$.

The cases in which α_c has been measured for growth of a thick film or crystal are summarized in Table 11. The earliest measurements are those of Volmer and Estermann[230] who found $\alpha_c \simeq 0.93$ for mercury at $-60°C$ and $-95°C$. This result is based on the assumption that $\alpha_c = 1$ at $-180°C$, as implied by Knudsen's observations.[361] Since they also observed that mercury crystals grew as platelets at $-60°C$, it is likely that the small apparent deviation from unity was due to the kinetics prescribed by equation (E-15) at the lowest temperatures. Further, they worked in a low vacuum and hence impurity adsorption was a possibility.

Haward[358] studied the growth of HgI_2 and found results consistent with $\alpha_c = 1$ for high supersaturations and of the form of Fig. 33 down to supersaturations approaching zero. However, in order to obtain this result he had to assume that the effective "equilibrium" vapor pressure was about 1.7 times the true equilibrium vapor pressure. He explained this seemingly strange result by supposing that the vapor condensed as metastable yellow orthorhombic phase, which must have a higher vapor pressure than the stable red tetragonal phase. Newkirk[362] studied the solid-state transformation of the HgI_2 and found that the surface-vapor interfacial free energy of the red phase is greater than that of the yellow phase by 15%. Hence it is reasonable that the yellow phase may nucleate and grow at a temperature where the red phase is stable, in agreement with Haward's hypothesis.

Hock and Neumann[357] studied the growth of polyhedral potassium crystals and found agreement with Fig. 33 with a $(p/p_e)_{crit}$ of 1.25, as shown in Table 11. However, they also observed whiskers growing from the polyhedral crystals, indicating that the crystals were imperfect and contained screw dislocations. Also, their observed value of $(p/p_e)_{crit}$ is appreciably less than the value calculated from equation (D-37). These factors, as discussed by Sears,[300] strongly suggest that impurity adsorption was present in their work. Hence equation (D-70) with $\alpha_{c_s} \ll 1$ describes their data for

TABLE 11

OBSERVED VALUES OF α_c, EXPERIMENTAL CONDITIONS

Values of $(p/p_e)_{\mathrm{crit}}$ calculated from equation (D-37) and parameters used in calculating $(p/p_e)_{\mathrm{crit}}$. Values in parentheses were not stated in the original references but are estimated by the present authors.

Material	Substrate temperature °C	Residual gas pressure, mm Hg	Beam flux, atoms/cm²-sec.	Super-saturation, ratio (p/p_e)	α_c	Ref.	Monolayer height, h, Å	$\Omega \times 10^{-23}$ cm³	σ ergs/cm²	$(p/p_e)_{\mathrm{crit}}$
Hg	−180 − 95 − 60	$(\sim 10^{-5})$	$\sim 5 \times 10^{16}$ $\sim 3 \times 10^{16}$ $\sim 3 \times 10^{16}$	$(\sim 10^{23})$ $(\sim 10^{6})$ $(\sim 10^{3})$	1.00 0.93 0.92	230	3.00	2.46	500	$\sim 10^{29}$ $\sim 10^{8}$ 3×10^{5}
Hg	to −133	10^{-10}	6×10^{13}	$\sim 10^{10}$	1.00	163	,,	,,	,,	$\sim 10^{12}$
Ag	20–500	$\sim 5 \times 10^{-6}$	? (small)	?	0.3–0.8	184	2.04	1.70	1140	$\sim 10^{7}$ to 11
Ag	≤ 300	$(\sim 10^{-5})$	$(\sim 10^{12})$	?	0.9–1.0	354	,,	,,	,,	74
Ag	−28 to 171	$\sim 10^{-5}$	$\sim 10^{16}$	$\gtrsim 10^{30}$ to 10^{18}	0.982 to 0.879	355	,,	,,	,,	$\sim 2 \times 10^{10}$ to 10^{3}
Ag	192	5×10^{-5}	10^{11}	$\sim 10^{12}$	1.00 ± 0.05	183	,,	,,	,,	10^{3}
Ag	45 200 200 440	10^{-8} 10^{-4} 6×10^{-10} 3×10^{-9}	10^{14} 4×10^{14} 4×10^{14} 10^{14}	$\sim 10^{31}$ $\sim 10^{16}$ $\sim 10^{16}$ $\sim 10^{6}$	0.99 ± 0.02 1.00 ± 0.01 1.00 ± 0.01 0.98 ± 0.02	356	,,	,,	,,	10^{6} 5×10^{2} 5×10^{2} 20

TABLE 11 (*continued*)

Material	Substrate temperature °C	Residual gas pressure, mm Hg	Beam flux, atoms/cm²-sec.	Super-saturation, ratio (p/p_e)	α_c	Ref.	Monolayer height, h, Å	$\Omega \times 10^{-23}$ cm³	σ ergs/cm²	$(p/p_e)_{crit}$
K	11	$(\sim 10^{-6})$	3×10^{14}	520	1.0	357	3.79	7.47	100	3.0
	58			1.4	0.95					2.2
	61			1.3	0.98					2.2
HgI_2	22	$(\lesssim 10^{-5})$	$\sim 2 \times 10^{16}$	10 to 100	~ 1	358			~ 100	
	14									
Cd	18	10^{-9}	2×10^{16}	2×10^7	0.97 ± 0.01	359	2.80	2.16	650	10^4
	19	5×10^{-7}	2×10^{16}	2×10^7	1.00 ± 0.01					
	23	10^{-4}	2×10^{16}	2×10^7	1.00 ± 0.01					
Zn	68	6×10^{-9}	10^{16}	8×10^6	0.96 ± 0.01	359	2.47	1.52	750	3×10^2
Au	800	$\sim 1 \times 10^{-6}$			> 0.99	360				
Rh	1200	$\sim 1 \times 10^{-6}$			> 0.99	360				
W	~ 600 to ~ 1900	$\sim 1 \times 10^{-6}$			0.998 ± 0.0005	360				
Pt	1200	$\sim 1 \times 10^{-6}$		10^4	> 0.998	360				
	300	$\sim 1 \times 10^{-6}$		10^{33}	> 0.998					
	600–800	$\sim 1 \times 10^{-6}$		—	~ 0.5					

$(p/p_e) < 1.25$; the supersaturation ratio 1.25 is interpreted as $(p/p_e)''$, equation (D-68).

Also Dittmar and Neumann[310] measured the growth rate of crystal planes on a potassium crystal containing low-index {110} planes, high-index {hkl} planes near the {110} planes and high-index {hkl} planes away from the {110} planes and found $\alpha_c \cong 0.9$, 1.15 and 1.0, respectively. They used a system similar to Hock and Neumann's with $(p/p_e) = 2.8$ and a substrate temperature of 59°C. Their vacuum may have differed from Hock and Neumann's but was probably $\lesssim 10^{-7}$ mm Hg. Since by the above arguments a (p/p_e) of 2.8 should have been greater than but near $(p/p_e)''$ (Fig. 36), their work indicates that (a) as expected, $\alpha_c = 1$ for high-index, roughened planes even under conditions when $\alpha_c < 1$ for low-index planes, and (b) atoms diffuse on low-index planes to an edge where they adhere to the high-index surface present there.

As this manuscript neared completion, Parker[363] communicated his results on the growth of polyhedral (110) surfaces of potassium. His experiment was quite similar to Hock and Neumann's except that he used potassium of higher purity and sealed off his growth tube after outgassing at 400°C and a pressure of 2×10^{-9} mm Hg. He found an α_c of unity for $(p/p_e) = 1.11$ to 4.1 at various temperatures near 340°K. Parker feels that dislocations were present in both his and Hock and Neumann's experiments and proposes that Hock and Neumann's observations of a low α_c for $(p/p_e) < 1.25$ may have been due to the effect of an impurity in stopping ledge motion, as proposed by Cabrera and Vermilyea.[276] Also he suggests that these observations are at odds with the theory of Hirth and Pound.[41] However, as pointed out by Hruska,[170] this alleged disagreement is probably non-existent because the fraction of perturbed surface area is quite large in the case of potassium,[326] even for low concentrations of dislocations. Thus the theory would predict an α_c of unity.

Extensive work has been carried out on the condensation of silver. Devienne,[354] using a radioactive-tracer technique and small beam fluxes, found that α_c rapidly approached unity as the condensing crystal grew thicker than about a monolayer. Fraunfelder,[184] also using a radioactive-tracer technique, a vacuum of $\sim 5 \times 10^{-6}$ mm Hg and a small but unreported beam flux, found α_c values of 0.3 to 0.6 for deposition of a few monolayers on presumably contaminated silver surfaces. Also, he observed α_c values of 0.4 to 0.8 on a freshly deposited film and therefore interpreted his results as illustrating the effect of impurities on α_c. In both the work of Devienne and of Fraunfelder, however, it seems probable that only a few silver nuclei were present on the substrate (see Section C) during growth and hence that a true measurement of α_c was not obtained.

Yang *et al.*,[183] also using a radioactive-tracer technique, observed $\alpha_c = 1$ for the deposition of a few monolayers on an initially outgassed silver crystal at 192°C in a vacuum of 5×10^{-5} mm Hg. Chandra and Scott[355] measured α_c by collecting on a target the atoms of silver reflected from a growing crystal. They found that α_c varied from 0.982 at $-28°C$ to 0.879 at $171°C$ in a moderate vacuum.

Because theory (equation D-37) predicts $\alpha_c = 1$ for a clean crystal, it was hypothesized[356] that contamination accounts for the observations of Chandra and Scott that $\alpha_c < 1$. Further it was thought that such contamination would not occur in high vacua of 10^{-9} to 10^{-10} mm Hg. Rapp et al.[356] performed an experiment similar to Chandra and Scott's, except that high vacua were used.* Also, the target, for collection of any silver atoms which might have been reflected from the growing silver crystal, was held at $-78°C$. Indeed it was found that $\alpha_c = 1$ in high vacua. However, it was also found that α_c equaled unity in a poor vacuum of 10^{-4} mm Hg where the ratio of silver to impurity-gas flux was 0.014.

Rapp et al.[359] also found $\alpha_c \cong 1$ for the condensation of Zn and Cd in high vacua and for the condensation of cadmium in vacua as poor as 10^{-4} mm Hg. In view of these results and the finding by Fraunfelder[184] and others, as reviewed in Section C, that gas-impurity contamination can lower α_c, Rapp et al.[359] concluded that $\alpha_c = 1$ for a clean metal with a monatomic vapor phase and that dynamic, adsorption of gas impurity may not lower α_c whereas static adsorption does lower α_c.

Stahl[364] studied thin films of various metals by X-ray diffraction after deposition in a vacuum of 10^{-4} mm Hg and found oxide patterns in films $\gtrsim 1$ micron thick for beryllium, aluminum, molybdenum, nickel, magnesium, calcium, and strontium. In thicker films only magnesium, calcium and strontium showed oxide patterns. Stahl's findings, together with the above results,[359] suggest that a critical ratio of impurity flux to flux of depositing material may exist above which dynamic impurity adsorption affects condensation kinetics.

Chandra and Scott[355] also found α_c to be less than unity for gold and copper in their experiments: for gold α_c varied from 0.95 at $157°C$ to 0.97 at $25°C$; for copper α_c varied from 0.80 at $157°C$ to 0.90 at $25°C$. However, on the basis of the above results we feel that impurity adsorption or some other factor led to the low measured values of α_c. For a clean surface it would be expected that α_c should equal unity.

Chupka[360] has performed a quite interesting set of experiments using a mass-spectrometer technique in which he could alternatively measure the evaporation flux from a source filament and the portion of this evaporation flux that did not condense on a second "reflecting" filament. He worked with several materials at various substrate temperatures as shown in Table 11. The residual pressure in Chupka's apparatus was $\sim 1 \times 10^{-6}$ mm Hg and consisted chiefly of H_2O, CO, and CO_2. However, the reflecting filament could be flashed at high temperature and then dropped in 1 to 2 sec to the experimental temperature. Since the mass spectrometer gave an essentially

---

* Of the order of 10^{-9} mm Hg. At a vacuum of 10^{-5} mm Hg the flux of residual gas striking a surface is ~ 1 monolayer/sec, so that contamination is always possible in such vacua.

instantaneous measurement of reflection (or re-evaporation), his values of α_c probably are those for clean surfaces. As may be seen in Table 11 Chupka found that $\alpha_c \simeq 1$ for gold, rhodium and tungsten, in agreement with expectation for clean metallic surfaces. In the case of platinum, α_c was about unity at 300°C and 1200°C but dropped to as low as 0.5 at 600–800°C. Also, when less than unity α_c was temperature and time dependent. When the temperature was lowered in 1 to 2 sec from ~1200°C to 700°C the reflection slowly rose during a period of $\sim\frac{1}{2}$ min. The fact that α_c was time-dependent could, we feel, indicate that $(p/p_e) \simeq (p/p_e)_1$ (Fig. 33) at 600–800°C and that the change in α_c was due to changes in $\bar{X}$ and λ^* from values characteristic of ~1200°C to those characteristic of ~700°C. Alternatively, time-dependent impurity adsorption may have affected the condensation kinetics at ~700°C. Chupka is continuing this interesting work. The advantages of using a mass spectrometer in which residual impurity gases can be identified as well as monitored are emphasized.

In the same apparatus described above, Chupka has measured $\alpha_c = 0.4 \pm 0.2$ for C atoms and $\alpha_c = 0.1 \pm 0.1$ for C_3 molecules impinging on spectroscopic graphite rods at ~2000°K and under supersaturations of 10 to 10^3. As the temperature of the substrate was lowered to 500°C there was some indication that C atoms combined to form C_3 on the surface. The small values of α_c found by Chupka for carbon cannot be interpreted in detail at this time. However, considering the fact that C_3 molecules have rotational degrees of freedom while C atoms do not, the finding that α_c is lower for C_3 molecules could be attributed to a δ effect. Even though data on α_c for graphite are sparse at this time, the results illustrate the complications that can be introduced when several polymers take part in the kinetics.

Knacke and Stranski[365] reviewed the results of Stranski and co-workers[366–368] on the growth of As, As_2O_3, P_2O_5, As_2S_2 and Sb_2O_3 where very small values of α_c ($\sim 10^{-7}$) were observed. This finding is attributed to the requirement for activated dissociation of vapor molecules before condensation can occur. For example, the equilibrium vapor over As_2O_3 consists largely of As_4O_6. In order for condensation of a vapor of As_4O_6 to occur, dissociation to As_2O_3 molecules must take place and this explains the low value of α_c in these cases.

With regard to the mechanism of growth at high supersaturations, it is observed that films grown from the vapor frequently have a preferred orientation with a low-index-plane pole normal to the surface (see the review by Bassett *et al.*[189]). This suggests that, even at supersaturations exceeding $(p/p_e)_{crit}$, growth proceeds by a two-dimensional nucleation process rather than by the advance of a diffuse interface.[261] However,[189] random orientations and/or recrystallized structures are also observed and this supports the diffuse-interface model. Thus while some evidence exists for transition to growth of a diffuse surface at high supersaturations, present evidence

* Equation (E-3).

suggests that the transition occurs only when $(p/p_e) > (p/p_e)_{crit}$. Hence growth rates are not affected by the onset of surface diffuseness but growth topography may be.

In summary, limited data for the growth of clean crystals with a monatomic vapor phase at high supersaturation indicate that $\alpha_c = 1$ in agreement with expectation. Contamination reduces α_c for such crystals. The presence of polymers in the equilibrium vapor over a crystal has been shown to lead to $\alpha_c \ll 1$ for some inorganic compounds.

2. *Topography*

(a) *Spirals.* Following the suggestion of Frank[324] that spiral dislocations account for crystal growth at supersaturations too low for growth by two-dimensional nucleation, a number of workers observed growth spirals, as reviewed by Frank,[338] Forty,[369] Verma[322] and Dekeyser and Amelinckx.[321] These observations on various materials of course tended to corroborate Frank's[324] view. Spirals have been observed, for growth from the vapor or from solution, on crystals of: beryl;[370] silicon carbide;[371,372] gold;[373] magnesium;[374] cadmium;[375,376] WO_3;[377] anthracene, fluorine, naphthalene, phthalinide, pyrine, carbazol, camphor and benzoic acid;[378] silver;[379] *p*-toluidine;[348,380] cadmium iodide;[381,382] paraffin ($C_{36}H_{74}$);[383] *n*-hectane;[384,385] and other crystals.[321,322] Most of these observations were on macroscopic spiral steps, which could be observed by means of interferometry, electron microscopy or light microscopy. In several instances,[370,371,374,376,379] however, ledges of monomolecular height were observed in phase-contrast microscopy on as-grown crystals where the ledges had been "decorated" by selective adsorption or compound formation.

Definite proof that spirals originate at spiral dislocation-surface intersections was provided by Frank and Forty in a study of silver growth.[379] They grew cubo-octahedrons of silver at 950°C with (p/p_e) probably $\gtrsim 1.01$ and subsequently decorated monatomic ledges on the crystals by contacting them with plasticine in air for 12 hours. They could distinguish ledges in spirals, which formed during growth at 950°C, and straight ledges due to slip of the dislocation subsequent to formation of the spiral (see Fig. 40). The finding that slip ledges terminated at spiral centers provided proof that the spiral had originated at a dislocation because the slip ledges could only be generated by dislocations.[379] Tanisaki[377] found similar results in the growth of WO_3. Anderson and Dawson[386] have shown growth spirals on *n*-nonatriacontane and stearic acid grown from solution, whose dislocations had slipped out of the crystal, as shown in Fig. 41.

Forty[381] and Newkirk[382] have observed double-dislocation sources (see Fig. 34) in operation in the growth of cadmium iodide from solution. Also Lemmlein and Dukova[387] and Lemmlein *et al.*[348] observed a double source in operation in the growth of *p*-toluidine from the vapor. Further, they found evidence for cases in which the two dislocations responsible for the

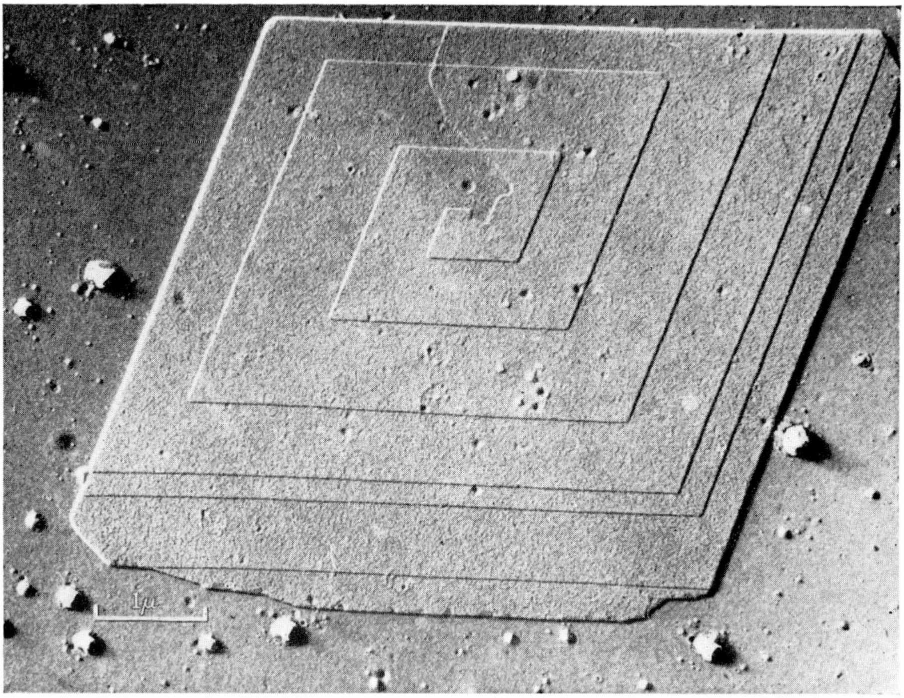

FIG. 41(a)

FIG. 41(b)

FIG. 41. Two examples of growth spirals on *n*-nonatriacontane[386] whose dislocations slipped out of the crystal after growth.

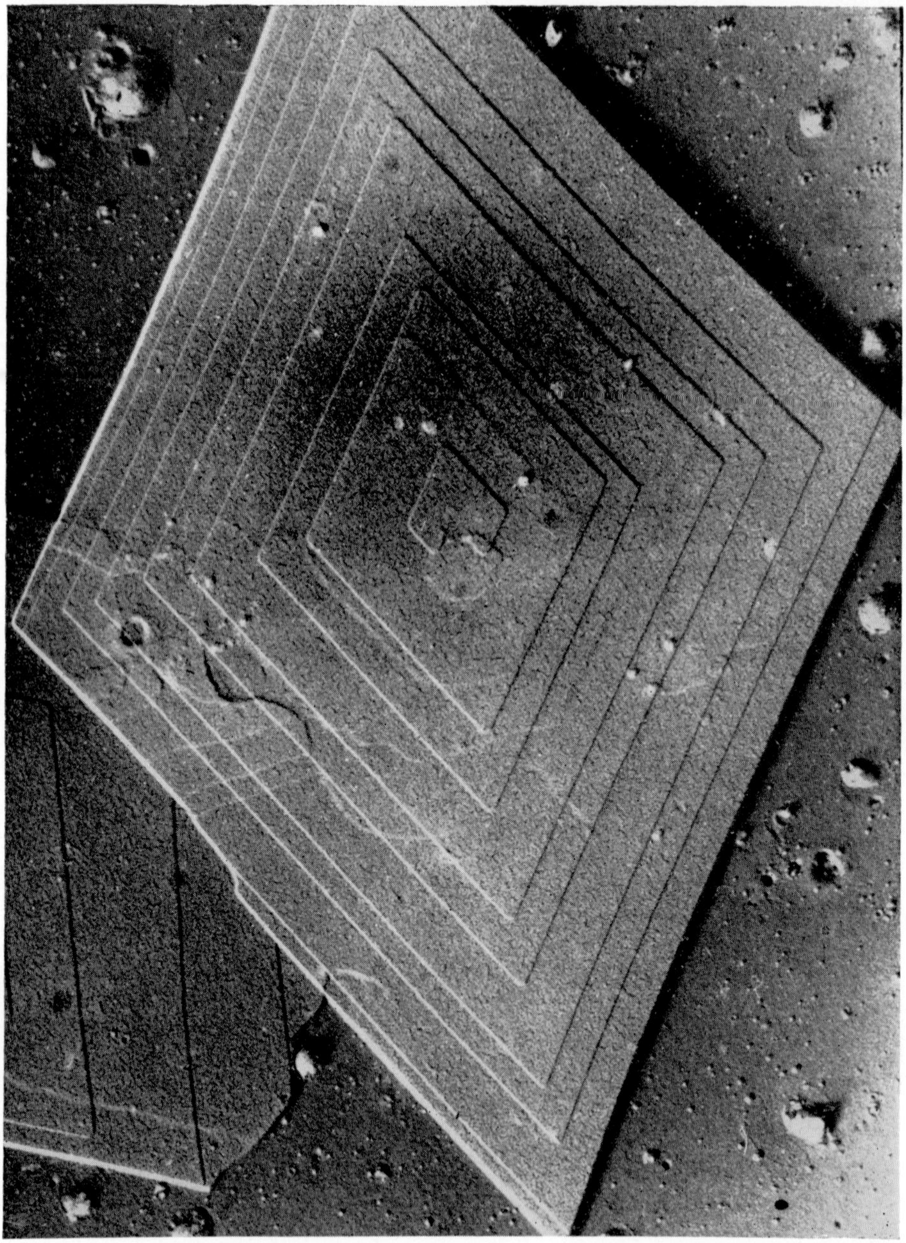

FIG. 42. Double-dislocation spiral source on *n*-nonatriacontane.[386]

source either slipped or grew together and hence mutually annihilated one another as growth centers. Lemmlein *et al.*[348] and Dukova[380] have observed in the growth of *p*-toluidine and of naphthalene that, upon changing the environment of a crystal from a supersaturated to an undersaturated vapor, the spiral changes over, beginning at the center, from a growth spiral to an evaporation spiral of the opposite sense. Anderson and Dawson[386]

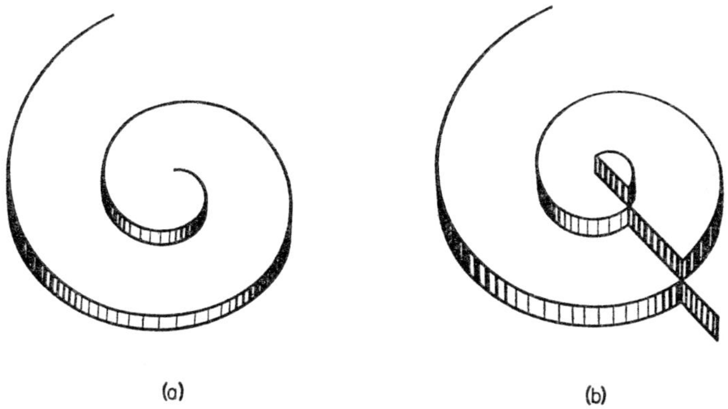

FIG. 40. Ledges on low-index silver surfaces[379] (a) as grown; (b) after slip.

have observed a double-dislocation source in the growth of *n*-nonatriacontane from solution as shown in Fig. 42.

Frank,[330] as noted above, suggested that dislocations of large Burgers vector give rise to polytypism in carborundum crystals. Forty[381] proposed the same view for the growth of cadmium iodide. He reasoned that ledges of odd step height should be less numerous than those of even step height because of the stacking fault associated with odd step heights. A statistical analysis of his data did show, in agreement with his hypothesis, that even step heights predominated.

Lemmlein and Dukova,[387] for *p*-toluidine and naphthalene, and Vlach,[388] for silicon carbide, have observed macroscopic growth spirals which broke up into spirals of smaller step height as the supersaturation increased, in agreement with the predictions of Cabrera[335] and of Chernov *et al.*[320,336]

In many of the above observations the spacing λ of ledges was constant, as would be predicted by either equation (E-4) or (E-6) for low supersaturation. However, in several instances[322,376,379] the spacing of the ledges increased with distance from the source, as would be predicted only by the argument[326] leading to equation (E-6). This finding can be interpreted as evidence for the non-steady-state behavior of a growth spiral in the region $\lambda_1 < \lambda_0$ (equations E-2 and E-6).

(b) *Straight ledges.* In studies of growth of polycrystals from the vapor by Newkirk[362] for HgI_2 and by Mehl and co-workers[347,375,389] for cadmium, macroscopic steps have been observed to originate near grain boundaries and then to propagate across the given single crystal. This was interpreted by these investigators as indicating that monomolecular ledges originated at the numerous spiral dislocations present at the grain boundaries and that these monoledges coagulated into a visible macroscopic step upon interaction with some "imperfection". In the light of Frank's[258] model of growth it is likely that the "imperfections" were impurities which had been adsorbed from the vapor.

Sears[316] has provided the clearest demonstration that grain boundaries can be multiple-dislocation sources in growth by a study of the growth of potassium chloride from dilute solution. He observed that single platelets nucleated at low supersaturations and grew only by edge-wise growth. However, when two platelets impinged upon one another, forming a twist boundary or crossed grid of screw dislocations at their junction, the platelets immediately commenced thickening. This was presumably due to the operation of the spiral dislocations at the twist boundary.* In another experiment, illustrating the role of dislocations in crystal growth, Korndorffer *et al.*[390] observed cadmium iodide platelets growing edgewise from solution but not thickening. If they indented such platelets on the presumably perfect basal faces, these faces immediately began thickening. This indicated that spiral dislocations were generated during indentation as would be expected from deformation theory.[266]

Similarly the observation of Sears and Coleman[391] that "butterfly wings" form when two zinc whiskers intersect in growth from the vapor provides confirmation of Frank's mechanism. Thus when the whiskers intersect, a twist boundary forms at their junction, providing spiral-dislocation sources which lead to the platelet or "butterfly wing" growth form. Price[288] has made a similar observation on intersecting cadmium whiskers.

(c) *Impurity effects.* The various observations on macroscopic steps cited above are interpreted as being due to impurity adsorption or possibly to stacking faults in some cases.[333] Forty[381] has shown directly that addition of CdI_2 "poison" increases the height of steps on lead iodide crystals growing from dilute solution.

Newkirk[382] (on cadmium iodide) and Forty[376] (on zinc) observed that incorporation of a macroscopic impurity particle into a growing crystal led to nucleation of concentric closed rings of steps at the particle until it was buried in the growing crystal. Observations on rods and platelets in growth from solution[316,382] have been interpreted[340] as confirming the view[339,340] that spiral dislocations are formed by incorporation of macroscopic impurity particles into a growing crystal. Kozlovski[392] has observed the formation of spiral dislocations by impurity capture in a platelet of β-methylnaphthalene growing from solution. He doped the solution with colloidal graphite particles and observed spirals emanating from points where graphite contacted the growing surface. He noted also that the ratio of the density of spiral dislocations

* This finding also corroborates the hypothesis in Section D-3 that basal faces of platelets can be regarded as perfect low-index surfaces.

on the growing surface to the concentration of graphite particles increased with increasing growth rate, indicating that the probability of a growth accident leading to spiral formation at an impurity was greater at higher supersaturations.

Sears(300) and others(313,363) have interpreted Hock and Neumann's(357) observation of whiskers growing out from a bulk potassium crystal in growth from the vapor as indicating impurity adsorption. Neumann and Dittmar(393) later interpreted their own observation on potassium whiskers in the same way. The explanation is that impurity adsorption must have occurred, leading to $d^2|J_c|/d(1/\lambda_c)^2 > 0$, which is Frank's condition(258) for a whisker to grow out of a bulk crystal when the mean free path in the vapor phase is large (see Part 2 of this Section).

(d) *Anomalous growth forms.* An unusual growth form has been observed by Kowarski,(394) Lemmlein *et al.*(348) and Yoda(395) and analyzed in some detail by the latter two groups of authors. In these investigations of growth from the vapor of *p*-toluidine(348) and MoO_3,(395) a liquid droplet was observed to form on an advancing macroscopic step. This causes the ledge to advance more rapidly in the vicinity of the droplet and leads to the growth topography indicated in Fig. 43. This phenomenon occurs under conditions

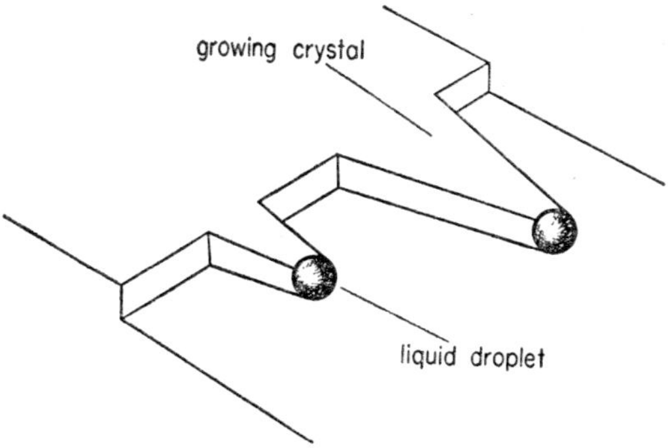

FIG. 43. Liquid droplets in contact with a macroscopic step, leading to more rapid growth in the vicinity of the droplets.

where the liquid is metastable with respect to the growing solid. The interpretation is that δ effects lead to an effective α_c such that the two-stage process vapor $\rightarrow$ liquid $\rightarrow$ solid is faster. The δ effects are thought to be due to the presence of polymers in the vapor phase that must dissociate to form monomers which are the fundamental units of the crystal growth.

In summary, topographical studies provide convincing evidence for spiral-dislocation growth at low supersaturation and for the surface-diffusion mechanism of crystal growth. The questions of whether ledge spacings approach an asymptotic value and whether they are specifically related to (p/p_e) or to $\bar{X}$ (equation E-4 or E-6) have not been unequivocally answered.

The role of impurities in macroscopic-step formation has only been indirectly demonstrated.

3. *Whisker and Platelet Growth*

Whisker growth in general has been the subject of a recent extensive review.[396] Whisker growth occurs by a number of possible mechanisms, including "extrusion" from the base, local reduction of a compound, electrolysis and other mechanisms. Here we are interested only in growth from the vapor phase involving adsorption, surface diffusion to the growing whisker tip and accretion there.

The first evidence for platelet growth was provided by Volmer and Estermann[230] who observed that, in growth from the vapor at supersaturations of $\sim 10^3$, mercury platelets grew edgewise at a rate 1000 times that expected from direct impingement. This demonstrated that growth must have occurred by surface diffusion on perfect basal surfaces to the edges where accretion to the growing crystal took place. Similarly, Sears[283] observed the growth of mercury whiskers at a rate 5000 times that expected from direct impingement. As discussed in Section (D-3), he also demonstrated that the growth took place at supersaturations less than the critical supersaturation for two-dimensional nucleation, a condition required for the stability of the low-index prism faces of the whiskers. Sears[282] later demonstrated that mercury-whisker growth followed the linear growth law of equation (E-14). In this work mercury whiskers nucleated and grew on a Pyrex substrate. Further, they appeared at the same sites in repeated growth tests. Thus his observations indicated that the whiskers inherited a spiral dislocation from the glass substrate.

Sears[169] demonstrated for cadmium, zinc, silver and CdS that whisker growth could occur only at supersaturations less than the critical supersaturation for two-dimensional nucleation. This finding confirmed the supposition that the whisker sides are perfect, stable, low-index crystal surfaces. Also confirming this view was the observation of DeVries and Sears[291] that, as Al_2O_3 whiskers grew from a region of low supersaturation at the tip to a region of high supersaturation, rapid growth stopped, the whisker tip started thickening and growth layers proceeded from the tip to the base. The onset of tip thickening presumably occurred when the supersaturation ratio at the tip reached $(p/p_e)_{crit}$. Similarly, Morelock[287] showed for chromium, nickel, iron, copper and gold that whisker growth occurred at supersaturations less than the critical supersaturation for two-dimensional nucleation. He found for all the metals but gold that he could prevent whisker growth by impurity poisoning if the flux of metal onto the substrate was less than that of residual impurity gas in his 10^{-6} mm Hg vacuum. With gold, which is less susceptible to impurity poisoning than the other metals, whiskers could be grown at fluxes as low as one-tenth the flux of impurity gas. Morelock found that thin whiskers, ~ 0.04 microns in diameter, grew to only ~ 2 microns in length while thicker whiskers, ~ 0.5 microns in diameter, grew to ~ 300

microns in length. This finding suggests that growth cessation occurred by slip of the spiral dislocations out of the whiskers, because in this mechanism it is predicted[342] that the thinner the whisker the more probable such slip becomes.

Coleman and Sears[397] grew zinc whiskers in the presence of an inert gas and demonstrated that "dendritic" growth of whiskers from cones on a zinc substrate occurs as predicted[329] (Fig. 35). Coleman and Cabrera[398] studied zinc and cadmium whiskers which had been grown in inert gas and found that the whisker prism faces were $(01\bar{1}0)$ and $(01\bar{1}1)$ planes, confirming the view that these faces must be perfect low-index surfaces. Similarly, Dittmar and Neumann[309] showed that, in growth from the vapor, potassium whiskers are bounded by $\{110\}$ prism faces.

Sears[16] confirmed Volmer and Estermann's[230] findings on the growth of Hg platelets from the vapor. In addition he showed that the platelet growth follows equation (E-15). He observed that, on a Pyrex glass substrate at $-63.5°C$ in a vacuum of $\sim 10^{-6}$ mm Hg, whiskers formed and grew at $(p/p_e) \simeq 100$, platelets formed and grew at $(p/p_e) \gtrsim 1600$ and the proportion of whiskers and platelets varied in the expected manner for $100 < (p/p_e) < 1600$. Although the growth of platelets or whiskers is thought to be understood, the reason that platelets should nucleate at one supersaturation and whiskers at another remains obscure. Indeed, Kobayashi[399] has found that, in various supersaturations from 1 to 50% and over a temperature range of 0 to $-30°C$, ice forms in the various morphologies of plates, needles, sheaths, columns, dendrites and hollow prisms.

Gomer[153] investigated the growth of mercury whiskers in a field-emission microscope at a tungsten-substrate temperature of $-78°C$ and a vacuum of $\sim 10^{-8}$ mm Hg for supersaturations of $\sim 10^2$. He found growth rates in agreement with equation (E-12) and observed growth in both limits corresponding to equations (E-13) and (E-14), respectively. He suggested that cessation of whisker growth occurred when two-dimensional nucleation commenced on the prism faces of the whiskers. However, he observed an Eshelby twist[400], which indicates the presence of a spiral dislocation, in only 20% of the whiskers after growth cessation. This means that growth cessation could have occurred by slip of the spiral dislocation out of the whisker.[342] Also it could have occurred by tip poisoning.[270] The Eshelby twist, indicating the presence of a spiral dislocation, has also been observed by Sears[401] in LiF whiskers grown from dilute aqueous solution and by Sears *et al.*[402] in Al_2O_3 whiskers grown from the vapor phase.

Melmed and Gomer[403] have extended Gomer's work and grown whiskers of nickel, copper, iron, platinum, titanium, gold, aluminum, barium, germanium, vanadium and silver, all of which had low-index axes and appeared to be bounded by low-index planes. The growth kinetics followed equation (E-12), again confirming the growth mechanism proposed by Sears.[169] Interestingly, they were able to grow whiskers of the high-temperature phases of iron and titanium at $\sim 300°K$. We suppose this to be the phenomenon discussed in Section C-2-a-7.* Parker and Hardy[404] have performed similar experiments with mercury and potassium whiskers. Gomer,[153] Melmed and Gomer[403] and Parker and Hardy[404] all determined surface-diffusion coefficients by the procedure discussed following equation (E-14). As there noted, we have reservations about this procedure, as do Parker and Hardy. Parker and Hardy found a large scatter in the D_s values which had been determined by the above procedure for potassium. They obtained a value of only 0.5 kcal for $\Delta H^{\star}_{sd}$ and a pre-exponential of $\sim 10^{-6}$ cm^2/sec, both of which seem too low. These unusual D_s values are attributed to the inadequacy of the method.

Several other interesting phenomena have been observed in whisker growth. Parker and Hardy[404] observed an effect of irradiation by light on the growth of potassium whiskers and associated this effect with photoemission of the potassium. Hoffman *et al.*[405] grew silver and gold whiskers by halide reduction in the presence of a field of $\sim$1000 volts/cm and found that whisker axes tend to follow the field lines. Spencer and Dragsdorf[406] grew potassium chloride whiskers from solution and annealed them at 550°C for two hours. They observed "branch" whiskers forming at regular intervals along the original whisker. Their interpretation of this phenomenon, which indeed seems the only likely interpretation, is that the spiral dislocation in the original whisker was part edge and part screw. Further, it is postulated that the spiral dislocation climbed into a spiral configuration. At each intersection of this spiral with the surface, a branch whisker formed at the resulting spiral-dislocation site. This finding confirms Webb's[337] hypothesis that spiral dislocations can climb out of a whisker.

Price[280] grew zinc platelets and whiskers by vapor diffusion through argon at $(p/p_e) = 1.5$ to 3 and $T = 400$°C. After slowly cooling these crystals, he viewed them by electron transmission microscopy. He also found that some platelets were free of dislocations while others were not. However, carefully mounted platelets were *always* dislocation free, and hence the dislocations in some crystals are thought to have been introduced during mounting in the specimen holder of the electron microscope. Similarly, he found cadmium platelets which had been grown from the vapor[288] and carefully handled[289] to be free from dislocations. Also Price[290] observed dislocation-free {111} platelets of platinum which had been grown from platinum vapor in air. Further, he found {111} dendrites of nickel bromide which had been grown from the vapor to be free of dislocations. Forty[407] used electron transmission microscopy to study lead iodide platelets which had been grown from solution and found that some platelets contained no dislocations parallel to the basal plane of the platelet while others did. These dislocations are thought to have been introduced near defects in the crystals during handling and not during the initiation of crystal growth.[408] Again, the absence of dislocations in some instances may be interpreted as being due to climb or slip of the

* Ostwald's "law of stages".

dislocations out of the platelets.* However, the possibility cannot be excluded that the platelets grew as perfect crystals, as discussed in the theoretical section.

Dittmar and Neumann[309] observed the growth of potassium whiskers at 59°C and $1 < (p/p_e) < 1.7$ in a vacuum estimated to be $\sim 10^{-6} - 10^{-7}$ mm Hg. They found growth rates in agreement with equation (E-11). However, they obtained a value for $\bar{X}$ from equation (E-11) and from this calculated a value for D_s which they felt was unreasonably high. Hruska and Hirth[313] pointed out that good agreement with theoretical expectation and a reasonable value of D_s could be obtained if one postulated impurity adsorption on the prism faces of the growing whiskers.

In summary, observed rates of whisker growth can be analyzed on the basis of existing theory. Again, the effect of adsorbed impurity on growth has been shown only indirectly. The conditions for whisker formation as opposed to platelet formation are not understood. Also, although several mechanisms for whisker-growth cessation have been proposed, the mechanism operative in particular instances of growth is moot. There is strong evidence in the cases where twists have been observed in whiskers that an axial spiral-dislocation mechanism accounts for growth. Indeed this is the most satisfactory of the various mechanisms proposed for whisker growth. However, in some cases of whisker growth and in many cases of platelet growth, the whiskers and platelets have been found to be dislocation free. The growth of dislocation-free crystals can be rationalized by supposing that they grew with dislocations present but that the dislocations subsequently slipped or climbed out. However, this latter evidence could also indicate that the platelets (and whiskers) grew as perfect crystallites. Further work is required to settle this important question.

3. Evaporation of Imperfect Crystals

(a) *Theory*

1. *Kinetics*

In evaporation, as discussed in Section D-3, ledges arise at crystal edges and thus are present on an evaporating crystal surface of low index whether spiral dislocations are there or not. In the part on evaporation of perfect crystals it was noted that the spacing of ledges near an edge source is small and decreases as evaporation proceeds, leading to an α_λ in equation (D-58) of unity in the region near the edge. Away from the edge source, on the other hand, the ledge spacing approaches λ_0 and $\alpha_\lambda \cong \frac{1}{3}$ from equation (D-67). At a spiral-dislocation source the ledge is constrained to terminate at the

* We refer again to the work of Anderson and Dawson[386] (see Fig. 41) and Lemmlein *et al.*[348,387] who actually observed slip of a spiral dislocation out of a platelet after growth had ceased.

dislocation. Just as in crystal growth, the ledge will wind up into a spiral. For a given spiral dislocation the sense of the evaporation spiral will be opposite to that of the growth spiral. The radius of curvature of the spiral is still given by equation (E-1) with ΔG_v from equation (D-47). In fact all of the spiral-dislocation kinetics in evaporation are exactly analogous to those for growth, and hence near a spiral-dislocation source the ledge spacing is given by equation (E-2).

At small undersaturations where $\lambda_1 \gg \lambda_0$, ledges of spacing λ_0, which emanate from edges, will dominate the kinetics. Even at high undersaturations where $\lambda_1 < \lambda_0$, the low-index surface will consist of ledges of spacing λ_0 except near the spiral or edge sources. Hence the principal effect* of imperfections in crystal evaporation will be to determine the total area near sources where $\alpha_\lambda \cong 1$. The effective value of α_λ to be used in equation (D-58) is[326]

$$\overline{\alpha_\lambda} \cong A_1 + A_2/3 \tag{E-16}$$

where A_1 is the fraction of the evaporating area near ledge sources and A_2 is the fraction where $\lambda \cong \lambda_0$ and nearly asymptotic ledge motion obtains. The affected area near spiral dislocations has been estimated[326] and is reported in Table 10. Other imperfections, such as cracks, pores, grain boundaries or macroscopic steps, would similarly have the effect of increasing A_1 in equation (E-16).

As discussed by Cabrera and co-workers,[325,409,410] the lattice-strain energy associated with a dislocation[266] provides an additional driving force for evaporation in the vicinity of the dislocation. This effect is opposite to that in crystal growth where the strain energy decreases the driving force. Because of this strain energy not only spiral dislocations but also dislocations with Burgers vectors lying in the plane of the surface can promote preferential evaporation. At such a dislocation the free energy of formation of a disc-shaped hole would be, instead of equation (D-29),

$$\Delta G^0 = 2\pi r h \sigma + \pi r^2 h \Delta G_v - (\mu b^2/4\pi\zeta)[\ln (r/a) - 1]. \tag{E-17}$$

Here μ is the shear modulus, b the magnitude of the Burgers vector and ζ is a factor equal to unity for a screw dislocation and one minus Poisson's ratio for an edge dislocation. Thus the free energy of formation will be less, the nucleation probability greater and the critical undersaturation less for nucleation of a hole at such a dislocation than on a perfect crystal surface. This strain energy effect is relatively important only for large b or small σ, or specifically[325,409,410] when

$$r_0 = (\mu b^2/8\pi^2\sigma) \gtrsim h. \tag{E-18}$$

Thus for metal crystals, where b usually does not exceed the unit cell dimensions, the strain energy effect is not likely to be important. For other crystals

* If the edges of a crystal are protected by impurity adsorption or by the dipole effect discussed by Stranski,[305] but the spiral sources are not, the spiral-dislocation sources will control the evaporation kinetics in the same fashion that they control the growth kinetics as discussed in the preceding section.

with larger values of b the effect may be appreciable. When condition (E-18) is satisfied the dislocation should open up into a hollow tube of radius r_0,[411] even with zero driving force, i.e. when $\Delta G_v = 0$.

When condition (E-18) is not satisfied, maximization of equation (E-17) yields the free energy of formation of the critical nucleus

$$\Delta G^\star = -(\pi\varepsilon^2/h\Delta G_v) - (\mu b^2 h/4\pi)\ln(-\varepsilon/bh\Delta G_v). \qquad \text{(E-19)}$$

The frequency factor for nucleation at a dislocation is identical to that of equation (D-51) except for a factor $X_\perp$, the fraction of surface sites intersected by dislocations. Thus $J_\perp$ the nucleation rate at dislocations is related to that on a perfect low-index surface (equation D-51) by the expression

$$(J_\perp/J) = -(X_\perp \varepsilon/bh\Delta G_v)\exp(\mu b^2 h/4\pi kT). \qquad \text{(E-20)}$$

In evaporation the steps from edges dominate the evaporation kinetics. Thus the nucleation at dislocations described by equation (E-20) causes only minor perturbations in the evaporation rate.

A final aspect in which evaporation differs from growth of crystals is the following. When evaporation ledges are traversing a crystal the minimum undersaturation ratio at a point midway between two ledges where $(\partial n_s/\partial x) = 0$ is, according to equation (D-56), given by[326]

$$(n_s/n_{s_e})_{\min} = \operatorname{sech}(\lambda_0/\sqrt{2}\bar{X}) \simeq 0.1. \qquad \text{(E-21)}$$

Hence unless $(p/p_e)_{\text{crit}}$ for evaporation of a perfect surface is greater than 0.1, the onset of two-dimensional nucleation will not occur even when $p = 0$. Thus as shown in Fig. 33, one expects $\alpha_{v_6} \simeq 1/3$ to hold down to $p = 0$ (or, when a more complicated form of equation (D-58) applies, $\alpha_\lambda \simeq \frac{1}{3}$ to hold down to $p = 0$).

The major interest here is in crystals for which only one molecular species participates in the evaporation or growth kinetics. Although the theoretical

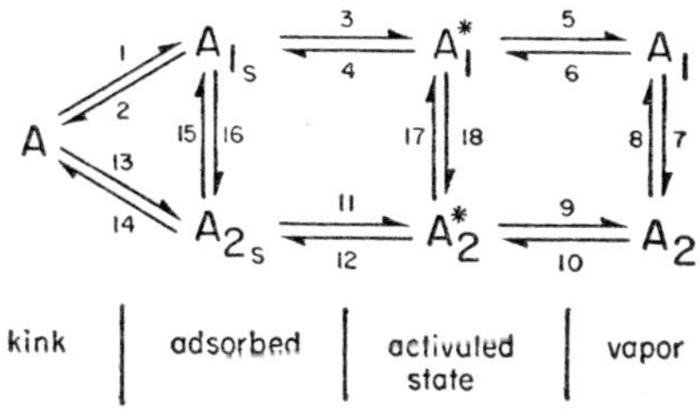

FIG. 44. Stages in evaporation in which both a monomer A_1 and a dimer A_2 participate in the reaction kinetics.

development leading to equation (D-58) applies only to such crystals, several other cases of evaporation kinetics are possible. If two polymeric species participate in the evaporation kinetics, the situation depicted in Fig. 44 arises; instead of three sets of reaction rates there are nine. The presence of more than two polymeric species would further complicate the kinetics. The general analysis of evaporation and growth kinetics in such a case has not been attempted. However, we note that if several polymers do participate in the kinetics δ (entropy) constraints are more likely for the higher polymers.

Also if vaporization occurs by decomposition of a compound to elements or by disproportionation to dissimilar molecules which are not both elements, a multiplicity of reactions similar to that shown in Fig. 44 is again possible.

In summary, evaporation kinetics should follow equation (D-58) and the ledge kinetics developed for evaporation of perfect crystals unless several polymeric species evaporate concurrently. The only effect of imperfections on evaporation should be to provide additional ledge sources.

2. *Topography*

Because steps from edges predominate over those from spiral dislocations, particularly at small undersaturations, spiral ledges should occur much less frequently on evaporated surfaces than on growth surfaces.

The effect of impurities on evaporation will be as discussed previously for growth. Impurity adsorption will promote macroscopic-step[(258)] and thermal-etch-pit formation. Such an etch pit simply represents a group of very closely spaced ledges from a dislocation source. There are at least two possible mechanisms for thermal-etch-pit formation. Impurities may adsorb at a ledge and lower σ and hence λ_1 (equation E-2) will be decreased.[(412)] Alternatively, if impurities adsorb in a time-dependent manner, they may lead to the formation of a macroscopic step around a dislocation source[(258)] of steps. Both of these effects would lead to visible thermal etch pits. Also if adsorbed impurity protects crystal edges, enhanced nucleation at dislocations, in accord with equation (E-20), may lead to pit formation even at dislocations which have no screw component normal to the surface.

(b) *Experiment*

1. *Kinetics*

(a) *Metal polycrystals.* Evaporation coefficients for polycrystals may be determined by comparing the results of Langmuir[(413)] and Knudsen[(273)] vapor-pressure measurements.* A compilation of α_v values determined in this way is presented in Part I of Table 12. These data refer to metals for which reliable results of both types of measurement are available. Consider the case of silver, for which Knudsen measurements of equilibrium pressure[(414,415)] are in excellent agreement. Jones *et al.*[(417)] present free-evaporation (Langmuir) data which yield an α_v value of 0.2. Similarly, Hirth and Pound[(418)] found α_v to be equal to 0.5 for polycrystals. Further, upon inspection of the surface the latter authors found that macroscopic steps had formed, presumably due to either impurity adsorption[(258)] or minimization

* See Section A on Coefficients of Condensation and Vaporization and Thermal Accommodation.

Table 12

Evaporation Coefficients for Various Materials

Material	Surface	Residual gas pressure, mm Hg	Surface temperature, °C	(p/p_e)	J_v $cm^{-2} sec^{-1}$	α_v	Ref.
(I)							
Ag	99.99 filings	$<10^{-5}$	860–951	K	—	1	414
Ag	99.99 filings	$(\sim 10^{-5})$	721–956	K	—	1	415
Ag	foil	$<10^{-5}$	960	0	$\sim 10^{18}$	1	416
Ag	wires	(10^{-3})	894–961	0	$\sim 10^{18}$	0.2	417
Ag	99.99 rod	10^{-5}	744–843	0	$\sim 10^{15}$	0.5 ± 0.2	418
Ag	filings	(10^{-4})	923–959	K	—	30% low	419
Cu	?	(10^{-5})	969–1067	K	—	1	420
Cu	filings	(10^{-5})	714–1084	K	—	50% low	421
Cu	wires	(10^{-3})	973–1025	0	$\sim 10^{15}$	0.4	417
Cu	99.98 sheet	(10^{-5})	995–1083	0	$\sim 10^{15}–10^{17}$	1	422
Cu	99.98 sheet	5×10^{-7}	873–1019	0	$\sim 10^{15}$	1	423
Be	99.9 powder,	$\sim 10^{-6}$	899–1282	K		1	231
	vacuum-cast, sintered			0	$\sim 10^{15}$	1	
	powder			0	$\sim 10^{15}$	0.6	
Be	99.9 sintered powder	$<10^{-5}$	900–1200	0	$\sim 10^{15}$	0.3	424
Be	99.9 sintered powder vacuum anneal	$\sim 10^{-6}$	904–956	0	$\sim 10^{15}$	0.8	425
	O_2 ads, N_2 ads	$>10^{-6}$	904–956	0	$\sim 10^{15}$	0.02 to 0.1 0.25	
Cd	99.95	(10^{-5})	208–261	K	—	1	426
Cd	99.9	(10^{-5})	249–303	K	—	1	427
Cd	contaminated with oil,	10^{-3}	198–234	0	$\sim 10^{18}$	0.1 to 0.0	428
	chemically cleaned			0		0.8+	
Fe	sheet	$<10^{-5}$	1282–1452	K	—	1	416
Fe	electrolytic filings	$<10^{-5}$	1227	K	—	50% low	429
Fe	electrolytic sheet	(10^{-5})	1144–1306	0	$\sim 10^{15}$	1	422
Fe	99.99 fused	10^{-6}	1181–1256	0	$\sim 10^{16}$	0.8	430
	deposit in vacuum					1.0	
Fe	99.97 vacuum cast	5×10^{-7}	1083–1246	0	$\sim 10^{15}$	0.5	431
Cr	99.9 filings	$<10^{-5}$	1100–1235	K	—	1	429
Cr	fused deposit					1.5	430
	in vacuum	10^{-6}	914–1237	0	$\sim 10^{15}$	2	
Cr	vacuum fused	10^{-6}	1010–1288	0	$\sim 10^{15}$	1	432
Cr	vacuum cast	10^{-6}	885–1015	0	$\sim 10^{13}$	0.5	433
Hg	deposit in vacuum	$(\sim 10^{-4})$	−39 to −64	0	$\sim 10^{14}$	~1	230
(II)							
K	deposit in vacuum	$<10^{-6}$	63	0	$\sim 10^{14}$	1.3	357
Cu, Ag	crystallites on W substrate	(10^{-5})	775–942	0	$\sim 10^{14}–10^{17}$	1.5 to 3.0	434
(III)							
Ag	(100), (111)	10^{-5}	744	0	$\sim 10^{14}$	0.4 ± 0.1	418
	(110)	10^{-5}	744				
			843	0	$\sim 10^{15}$	0.6 ± 0.2	
						0.8 ± 0.2	
KCl	(100), (111)	$\sim 10^{-5}$	439–547	$1>(p/p_e)>0$	$\sim 10^{14}$	0.7	435
	powder	$\sim 10^{-5}$		K		1	
	(110)	$\sim 10^{-5}$			$\sim 10^{14}$	0.7	
KCl	(100)	$\sim 10^{-5}$	477–597	0–0.8	$\sim 10^{16}$	0.5	436
KI	(100)		427–527	0–0.8	$\sim 10^{16}$	0.5	
Rhombic sulfur	(111)	$\sim 10^{-5}$	17–34	0	$\sim 10^{14}$	0.7	437
Rhombic sulfur	(111)	$\sim 10^{-5}$	20–40	0	$\sim 10^{14}$	0.61–0.65	438
	(011)					0.65–0.82	
	(001)					0.39–0.57	
	(113)					0.59–0.55	
CsI	low-index facets	$<10^{-5}$	484–499	0	$\sim 10^{17}$	0.3 ± 0.1	439
CsBr	low-index facets		512–557	0	$\sim 10^{17}$	0.3 ± 0.1	
NaCl	(001)		601–662	0	$\sim 10^{17}$	0.16–0.23	
Na_2Cl_2	(001)		601–662	0	$\sim 10^{16}$	0.11–0.18	
LiF	(001)		707–797	0	$\sim 10^{17}$	0.5	
Li_2F_2	(001)		707–797	0	$\sim 10^{17}$	0.6	
Li_3F_3	(001)		707–797	0	$\sim 10^{16}$	0.1–1	

TABLE 12 (*continued*)

Material	Surface	Residual gas pressure, mm Hg	Surface temperature, °C	(p/p_e)	J_v cm^{-2} sec^{-1}	α_v	Ref.
Rhombic sulfur	(111)	5×10^{-6}	60	0.6–1	$\sim 10^{16}$	0.6	344
Benzophenone	(110)	5×10^{-6}	20–30	0.75–1	$\sim 10^{16}$	~0.2	344
	(111)		20–40	0.6–1		0.4–0.7	
	polycrystal		15–30	0.85–1		0.7	
(IV)							
Ice		$<10^{-3}$	−57	0.95	$\sim 10^{18}$	1 ± 0.5	440
Ice		10^{-5}	−85 to −60	0	$\sim 10^{14}$	0.94	441
Ice		—	—	0	—	0.07	442
$HgCl_2$				0		~0.6	443
$HgBr_2$				0		~0.6	
HgI_2				0		~0.5	
Te_2		$<10^{-4}$	317–407	0	$\sim 10^{17}$	0.4	444
Se_2, Se_6			176	0	$\sim 10^{18}$	~1	
As_2, As_4			260	0	$\sim 10^{15}$	0.05	
P_2, P_4		$<10^{-4}$	305–408	0	$\sim 10^{15}$	10^{-7}	445
P_4		10^{-6}	250–450	0	$\sim 10^{15}$	10^{-7}	446, 447
NH_4Cl		10^{-4}–10^{-5}	118–221	0	$\sim 10^{14}$	0.003–0.0004	448
Al_2O_3		$<10^{-5}$	1600	0	—	2×10^{-4}	449
BN		10^{-6}	1220–1720	0	—	10^{-3}	450
AlN			1220–1720	0	—	10^{-8}	
C		$<10^{-6}$	2123–2233	0	$\sim 10^{15}$	0.37	451
C_2						0.34	
C_3						0.08	

Values listed in parentheses were not cited in the original references but are estimated by the present authors. "K" in (p/p_e) column signifies a Knudsen,[273] near-equilibrium, vapor-pressure measurement. When more than one Knudsen measurement is cited, one measurement is taken to yield the true equilibrium vapor pressure and the others are listed as −% high or low.

of surface free energy.[452] Because these macroscopic steps or facets were partially constituted of {111} planes, it was proposed that the apparent value of α_v for a polycrystal arose from actual values of $\alpha_v \cong \frac{1}{3}$ for the {111} regions and unity for the high-index areas of the facets. However, Wessel[416] performed an experiment in which two holes were drilled near the ends but on opposite sides of a hollow silver cylinder that was mounted as a torsion pendulum. Thus if the Knudsen equilibrium flux from the holes exceeded the free evaporation flux from the cylinder surface a deflection of the pendulum would be observed. He showed by this method that $\alpha_v > 0.92$ at 960°C where impurity adsorption is less likely than at lower temperatures. Hence the available results for polycrystalline silver suggest that $\alpha_{v_6} \cong 1$. This is in agreement with equation (D-61) because on the high-index planes of a random surface the λ values should be very small (approaching the lattice parameter). The free-evaporation[417,418] values of α_v less than unity are thought to be due to impurity adsorption in the poor vacua attained in these experiments.

The results for beryllium[231,424,425] show the effects of impurity on α_v. Sintered powder exhibited evaporation coefficients of 0.3[424] and 0.6,[231] while for vacuum-cast beryllium α_v was found to be unity. Gulbransen and Andrew found that the residual pressure in their system had to be $\gtrsim 10^{-6}$ mm Hg in order to avoid contamination effects. Oxygen chemisorption reduced α_v to 0.02–0.10 and nitrogen reduced α_v to 0.2. Thus it is thought that the kinetics may have followed equation (D-70). Surface topographies were not reported so it is not known whether macroscopic steps formed or not.

Bennewitz[428] observed that α_v was $\sim$0.01 for a contaminated cadmium surface but $\sim$1 for a clean surface. He stated that $\alpha_v \cong 0.5$ for a clean surface; however the equilibrium pressure which he used is $\sim$50% higher than the more reliable value quoted in Table 12.

The equilibrium vapor pressures of both iron and chromium are known only to $\pm$50%. Within this limit, $\alpha_v \sim 1$ for these metals. Higher purity chromium gave higher values of α_v,[429] again illustrating impurity effects.

Only cases where both Langmuir and Knudsen measurements are available are cited in the table for the following reasons. In Langmuir measurements the value of α_v is sometimes estimated to be unity if the calculated third-law values of ΔH_0^0 are constant for the range of temperatures studied (see e.g. ref. 422). However, ΔH_0^0 is a function of $RT \ln \alpha_v$ plus a constant. Hence over the limited temperature range of most measurements the third law test is useful only to the extent of showing that $\alpha_v \gtrsim 0.1$. Also, unless several independent determinations of a vapor pressure have been performed, the uncertainty in a single determination must be placed at about 50%.*

Carpenter and Mair[454] cite measurements on the evaporation of gold filaments (99.95 or 99.999%) in vacua of $\sim 10^{-5}$ mm Hg at 850–950°C in which the free evaporation rate was such that $\alpha_v \sim 0.1$. On the other hand, at 3×10^{-3} mm of oxygen pressure it was found that $\alpha_v \cong 1$. They interpreted these results by assuming that the oxygen reacted with the impurity on the gold and removed it as a volatile oxide. In their work, however, they estimated the rate of evaporation by observing the decrease in intensity of light transmitted through a cool glass plate on which the gold condensed and measured the evaporation rate directly in only one instance. Hence it is felt that the oxygen may have affected the rate of deposition of gold onto the cold plate rather than the rate of evaporation of the filament. It would be interesting to check these possibilities in a free-evaporation measurement. In an analogous study of silver evaporation, Ernas and Shewmon[455] found that the evaporation rate of silver at 900 and 975°C was unaffected by oxygen mole fractions in the vapor of 0, 0.2, 0.4, 0.6, 0.8 and 1.0 in a vapor phase of total pressure 10^{-1} mm Hg in which the balance of the vapor was nitrogen.

In summary, for the metals listed in Part I of Table 12 the proportion of polymers in the equilibrium vapor is negligible[456] and hence δ (entropy) effects of polymers should be absent. Thus, in agreement with equation (D-61), available evidence indicates that $\alpha_{v_6} \cong 1$ for clean high-index surfaces of metal crystals. Deviations from equation (D-61) are likely to be the result of impurity adsorption, either directly, or indirectly through the formation of macroscopic steps.

* Winterbottom and Hirth[453] have recently shown that part of this uncertainty is attributable to the neglect of a surface-diffusion contribution to the Knudsen-cell effusion flux.

The experiments listed in Part II of Table 12 yield anomalously high vaporization coefficients. It is suggested by the present authors that this effect was due to surface diffusion on an inert substrate. The potassium specimens were small crystallites which had been deposited on a silver wire; the copper and silver specimens were thin films on a tungsten substrate. Taking the case of potassium as an example, surface diffusion away from the crystallite on the silver wire and subsequent evaporation could have led to an apparently high value of α_v. This effect has been clearly demonstrated by Clancey,[(457)] who showed that the apparent evaporation rate of naphthalene was enhanced by surface diffusion on various substrates. Indeed, the experiments on whisker evaporation by Dittmar and Neumann[(309)] showed the same effect; the apparent evaporation rate of potassium whisker tips was 1000 times the maximum rate expected for direct evaporation from the tips because of surface diffusion down the prism faces of the whiskers. Of course, high evaporation rates may also be caused by oxidation in poor vacua under conditions such that the oxide is volatile.

(b) *Single crystals.* Evaporation data for various single crystals are listed in Part III of Table 12. For low-index (100) and (111) surfaces of silver crystals, Hirth and Pound[(418)] found that $\alpha_v = 0.4$ in close agreement with equation (D-67). However, considering their poor vacuum of 10^{-5} mm Hg, the possibility of impurity adsorption (equation D-70) cannot be excluded. Nevertheless, examination of the surface indicated that no macroscopic steps had formed during evaporation. For (110) surfaces of silver, α_v was apparently between 0.6 and 0.8. However, the (110) surfaces formed macroscopic steps containing segments of {hkl} and {111} planes, and this circumstance could have yielded α_v values intermediate between 0.4 and 1, in agreement with observation.

Bradley and Volans[(435)] studied the vaporization of (100) surfaces of potassium chloride. They found that $\alpha_v = 0.7$, which Bradley[(458)] has interpreted as $\alpha_{v_7} = \delta_{\text{ad}}$ (equation D-63). However, the result could equally well be interpreted as $\alpha_v = \alpha_\lambda = \alpha_{v_6}$ (equation D-61). Knacke *et al.*[(436)] found that $\alpha_v \simeq 0.5$ for (100) surfaces of both potassium chloride and potassium iodide. They placed their specimens at the bottom of silver tubes which had length to diameter ratios of from 1.4 to 25 and thus it was necessary to use Clausing's[(459)] correction factor for the tube resistance to vapor transport. The present authors note a reservation with respect to these results in that Clausing's factor does not take into account the surface diffusion in the tube.[(453)] In this connection it seems anomalous that α_v should decrease with increasing length/diameter ratio.* Finally, the work of Miller and Kusch[(460)] indicates that ~10% dimer is present in the vapor over potassium chloride and this could also have an effect on α_v.

* Later work by Stranski and Hirschwald[(530)] indicates that the macroscopic-step spacings were about 200 Å, somewhat less than λ_0, at large undersaturations (short tube lengths). On the other hand, at small undersaturations the macroscopic-step spacings were about 1 micron, appreciably greater than λ_0. Thus the apparent anomaly is resolved, and their results are consistent with the limiting-law model of Hirth and Pound.

Rhombic sulfur single crystals have been studied by several authors[344,437,438] who are in good agreement that $\alpha_v = 0.6$–0.7 for (111) surfaces. This result has been interpreted[223,437] as $\alpha_{v_7} = \delta_{ad}$ (equation D-63). However, again the results could be interpreted as $\alpha_v = \alpha_\lambda = \alpha_{v_6}$ (equation D-61). There is some support for the latter view in Fig. 37. It is seen that for the (110) benzophenone and the (111) sulfur surface the evaporation rate is higher than the growth rate at an equivalent supersaturation and that this difference is greater at larger undersaturations. This is the effect one would expect because growth does not bring about change in local orientation of the low-index surface while evaporation does, with accompanying decrease in ledge spacing λ. At higher undersaturations evaporation will proceed to a greater extent in a given time and a greater fraction* of the surface will be covered with ledges of smaller spacing. Further, curve (b) shows a constant $\alpha_v \cong 0.8$ for the polycrystal of benzophenone, which is what one would expect for {hkl} planes with ledge spacings $\lambda \ll \lambda_0$. Hence we feel that while equation (D-63) may apply to these results with $\delta_{ad} < 1$ there is at least some contribution of surface diffusion giving $\alpha_\lambda < 1$.

Rothberg *et al.*[439] studied the evaporation of several halides and demonstrated the effect on α_v of polymers in the vapor phase. For cesium iodide and cesium bromide polycrystals, which evaporated as monomers, they found that $\alpha_v = 0.3$. They kindly sent their specimens to the present authors who observed that the surfaces had formed low-index planes in macroscopic steps or facets. Hence these results may be interpreted as consistent with equation (D-67) and an $\alpha_{v_6} \cong \frac{1}{3}$. Also, Rothberg *et al.* found that $\alpha_v = 18.7 \exp(-8200/RT)$ for sodium chloride and that $\alpha_v = 33.6 \exp(-9800/RT)$ for Na_2Cl_2. The temperature dependence of α_v when several polymers are evaporating simultaneously is not surprising because of the association and dissociation reactions that may occur. These reactions may take place (i) in removal of a molecule from a kink position, (ii) in the adsorbed state and (iii) in the process of desorption to the vapor.

In summary, results for single crystals may be interpreted as indicating surface-diffusion or entropy constraints. Distinction between the two can be provided by considering surface topographies and comparing results for polycrystals and single crystals. In one instance, for benzophenone, such a comparison indicated that both constraints contributed to α_v. Further work like the latter would clarify the role of δ effects, particularly if carried out in high vacua. Most work to date has been accomplished in poor vacua and hence impurity adsorption effects cannot be excluded.

(c) *Polycrystals with polyatomic vapor species.* Results on polycrystals of materials other than metals with monatomic vapors are listed in Part IV of Table 12. Early results on ice[442] suggested that $\alpha_v \cong \alpha_{v_7} \cong \delta_{ad} = 0.07$. This is the behavior one would expect if the interpretation of the low vaporization coefficient of water as an entropy effect is correct (see Table 8). Kramers

* Corresponding to high-index planes.

and Stemerding[440] obtained a value of $\alpha_v \simeq 1$, but they worked under conditions where diffusion constraints in the vapor phase were important. Thus their result could be higher than the true free-evaporation value. However, Tschudin[441] found that $\alpha_v \simeq 1$ in free evaporation. If indeed $\alpha_v \simeq 1$ for the random {hkl} surfaces of a polycrystal, this would suggest that $\alpha_v = \alpha_{v_6} = \alpha_\lambda \simeq 1$, i.e. that no entropy ($\delta$) effects are present. Also it suggests that δ effects are not important in water evaporation. In view of the work on water evaporation cited in Table 8, further research to define the role of δ effects in either ice or water evaporation would seem to be in order.

Metzger and Miescher[443,444,461] found that $\alpha_v \simeq 0.6$ for several mercuric halides. In later work[444] with a torsion cell, in which both the evaporation rate and evaporation pressure were measured, it appeared that $HgCl_2$ evaporated as a monomer. Hence for $HgCl_2$ the evaporation coefficient of 0.6 is interpreted as a δ effect. If one supposes by analogy that $HgBr_2$ and HgI_2 also evaporate as monomers, their evaporation coefficients could be interpreted in the same manner. In addition[444] it was found that tellurium evaporated as a dimer with $\alpha_v \simeq 0.4$ and selenium as 90% dimer and 10% hexamer with $\alpha_v \simeq 1$.

Melville and Gray[445] found that $\alpha_v \simeq 10^{-7}$ for red phosphorus. They suggested that red phosphorus evaporates as P_2, while the equilibrium vapor consists chiefly of P_4. Referring to Fig. 44 (where for A_1 read P_2 and A_2 read P_4), they postulated that rates 9 and 10 are small due to a small $\alpha_c \simeq 0$ for P_4 and hence are negligible compared with rates 7, 8, 5 and 6. Thus in free evaporation rate 5 would be controlling. However, even if α_v for P_2 were unity, the overall α_v would be about 10^{-7} because the mole fraction of P_2 in the vapor is of the order of 10^{-7}. More recently, Kane[446] studied red phosphorus and found that $\alpha_v \simeq 10^{-7}$. However, Kane and Reynolds[447] determined by mass spectrometry that both the equilibrium vapor and the free-evaporation flux from red phosphorus consist of only P_4. In the light of this evidence Melville and Gray's view cannot hold, and red phosphorus must evaporate as P_4 with an $\alpha_v \simeq 10^{-7}$. The low value of α_v for P_4 may be interpreted according to equation (D-64) as due to an activation energy for movement of a P_4 molecule from a kink position to the adsorbed state. This is consistent with the view of Pauling and Simonetta[462] that one covalent bond per molecule must be broken in such a movement.

Metzger,[444] using a torsion apparatus, found arsenic to evaporate as 70% As_2 and 30% As_4 with an α_v of about 0.05. The low value of α_v was interpreted in the manner of Melville and Gray.[445] However, according to Kane and Reynolds[447] only As_4 is present in free evaporation, and thus the mechanism of evaporation of arsenic is moot also. While the interpretation of α_v for red phosphorus and arsenic is still somewhat controversial, the above discussion at least illustrates the complexities introduced into the evaporation kinetics when several polymers participate in the reactions. Briefly mentioning other examples of cases of complex evaporation, Chupka and Inghram[463,464] showed by mass spectrometry that the equilibrium vapor over graphite consists of C_3, C_1 and C_2. Thorn and Winslow[451] measured an overall value of $\alpha_v \simeq 0.15$ for graphite. Using Chupka and Inghram's[463] values for relative ion currents of the three species in free evaporation, Thorn and Winslow estimated ionization cross-sections for the three species and calculated $\alpha_v = 0.08$, 0.37 and 0.34 for C_3, C_1 and C_2, respectively. Dreger *et al.*[450] found that $\alpha_v = 10^{-3}$ for AlN and BN, which undergo decomposition to Al(g) and N_2(g) and B(s) and N_2(g), respectively, in free evaporation. Further, they observed that α_v is temperature dependent, indicating an activation energy for evaporation. Sears and Navais[449] found that $\alpha_v = 2 \times 10^{-4}$ for alumina, which disproportionates to Al, O, AlO, Al_2O and Al_2O_2 on vaporization.[464] Spingler[448] observed that $\alpha_v \simeq 10^{-7}$ for NH_4Cl, which dissociates to NH_3 and HCl on evaporation. However, he found an activation energy for evaporation of 13 kcal, in contrast to an equilibrium heat of vaporization of 39 kcal. He interpreted these results as consistent with equation (D-64). Detailed consideration of these kinetics will not be undertaken by the present authors, but it is noted that the complications of Fig. 44 may be present.

In summary, complex kinetics associated with evaporation of polymers or compounds have been illustrated. At present the kinetics of such reactions cannot be analyzed unequivocally because of insufficient data on the identity of the evaporating species as a function of time, temperature, undersaturation and crystal morphology.* In particular, results for single crystals in high vacua would be highly desirable. The burgeoning field of mass spectrometry[456] shows promise of providing such data.

2. *Topography*

(a) *Spirals and steps.* Lemmlein *et al.*[348] and Dukova[380] have observed spirals of polymolecular height in the growth of *p*-toluidine and of naphthalene and have observed that, upon changing the environment of a crystal from a supersaturated to an undersaturated vapor, the spiral changes over, beginning at the center, from a growth spiral to an evaporation spiral of the opposite sense. Dukova[465] has studied further the evaporation of *p*-toluidine and naphthalene. He found that the velocity of evaporation steps increases and the spiral spacing decreases with increase in supersaturation, in agreement with equations (D-66) and (D-69). Also he observed that the polymolecular steps break up into smaller ledges at high undersaturations. In view of the probable low vacuum in his work, we interpret this finding to mean that the polymolecular steps form by Frank's[258] dynamic adsorption mechanism. Further, at high undersaturations the step velocity becomes so great that Frank's mechanism no longer obtains. Then, as postulated by Hirth and Pound,[262] the polymolecular steps break up by acting as sources for smaller ledges.

* However, a very recent review by Stranski and Hirschwald[530] cites new data for evaporation of complex species, and these data are in good agreement with the theory developed by the Stranski school.

Votava and Amelinckx[466,467] heated sodium chloride crystals to 780°C for four days in a vacuum and then decorated the crystals by exposing them to slightly moist air for several days. They observed evaporation spirals which were several lattice spacings high on (001) surfaces. In one instance they found a row of spirals of monomolecular height originating at a sub-boundary. The latter finding was thought to demonstrate that the spirals had emanated from spiral dislocations. Bethge and Schaffer[468] grew copper crystals of 99.99% purity from the melt in a vacuum of 10^{-5} to 10^{-6} mm Hg and observed macroscopic evaporation spirals on (111) planes. Votava and Berghezan[469,470] heated copper in a vacuum of about 10^{-4} mm Hg at 900–1000°C and found macroscopic spirals emanating from incoherent twin boundaries, again indicating that the spirals had originated at spiral dislocations. Suzuki[471] heated (100) and (111) surfaces of copper to within 1°C of the melting point in a vacuum of 10^{-3} mm Hg and observed spiral steps which were supposed to be decorated by oxide particles. He also observed, analogously to the findings of Forty and Frank,[379] evidence for slip steps. Hence he concluded that the spirals had originated at spiral dislocations. Young[472] found macroscopic spirals around (111) poles of a copper sphere which had been heated to 1050°C in a vacuum of 5×10^{-9} mm Hg. Thus in all of these cases spiral ledges are associated with spiral dislocations. In cases where macroscopic steps were formed it is probable that impurity adsorption, either from the vapor phase in the poor vacua or from the original solid phase, led to macroscopic-step formation by Frank's[258] mechanism.

Macroscopic steps were observed[418] to form on {hkl} planes of polycrystals of silver during evaporation at high fluxes, and these steps were shown to affect the evaporation kinetics.[418] In this case it is likely that Frank's[258] impurity mechanism led to the formation of the macroscopic steps. Similarly, macroscopic steps have been observed on Cu after evaporation in high vacua.[472] Presumably these steps were caused by adsorption of an impurity from within the crystal. Also, macroscopic steps have been found on Cu after evaporation in poor vacua.[470]

Macroscopic steps have also been observed on chromium which had been heated to 1300–1500°C in 1 atm of H_2,[473] on silver which had been heated to 800–900°C in 1 atm of air or oxygen[474–477] and on copper which had been heated in air at 0.5 mm Hg.[478] All of these experiments were performed under conditions where diffusion through the vapor phase could have been rate controlling, a situation which would lead to a near-equilibrium boundary condition at the surface. Hence the results may indicate macroscopic-step formation by either Frank's[258] kinetic mechanism or equilibrium faceting.[452] The absence of macroscopic steps on silver which had been heated in air at 1 atm in a silver enclosure where no evaporation could have taken place[477] favors the view that the kinetic mechanism[258] led to macroscopic-step formation, at least in the case of silver.

Young and Gwathmey[478] performed an interesting experiment in which they heated spherical single crystals of copper to 900–1080°C in a high vacuum of 0.5 to 1 × 10^{-8} mm Hg. 99.98% purity copper exhibited macroscopic steps with low-index facets around (100), (110) and (111) poles and macroscopic spirals around (111) poles. 99.98% copper which had been heated in a copper cup to approximately equilibrium conditions showed shallower facets around the low-index poles and also etch pits of densities from 10 to 10^7 cm^{-2} about these poles. These observations suggest that impurities within the copper accumulated and adsorbed on the surfaces during evaporation, causing macroscopic-step formation. On the other hand, 99.999% copper exhibited only very faint macroscopic steps when heated in vacuum or in a 99.999% copper cup. However, if heated in a 99.98% copper cup, the 99.999% copper formed facets similar to the 99.98% copper. Again, this illustrates the role of impurities in causing macroscopic-step formation.

In summary, spiral steps associated with spiral dislocations have been observed on evaporated low-index surfaces. Observations of macroscopic steps under conditions of rapid evaporation have been related to impurity adsorption, either from the vapor or from within the evaporating crystal. Under small undersaturations macroscopic steps may arise kinetically by an impurity-adsorption mechanism, or they may arise because of equilibrium faceting. Further work appears necessary to clarify the mechanism of facet formation at small undersaturations.

(b) *Etch pits.* Thermal etch pits* have been observed to form under conditions where no oxidation reaction was proceeding on a number of materials including nickel, zinc, copper,[480] chromium,[473] sodium chloride, potassium chloride, potassium bromide[481] and titanium.[482] Hendrickson and Machlin[483] bent single crystals of silver and thermally etched them at 600°C in an argon–10% oxygen mixture at one atmosphere. They found that the etch-pit density after bending and etching corresponded to the dislocation density expected from the plastic deformation. We have already mentioned Young and Gwathmey's study[478] which also revealed that etch pits, presumably due to dislocations, formed on low-index surfaces of impure copper only under near-equilibrium conditions. This is consistent with expectation because, on a clean crystal under conditions close to equilibrium, ledge spacings around dislocations should be so large that microscopically visible pits should not form.[411] Rather they should form on surfaces that (i) contain adsorbed impurities, which give small ledge spacings near the dislocation sources,[258] and (ii) are exposed to a very small undersaturation, which prevents ledges that arise at crystal edges from sweeping over and obscuring the dislocation etch pits.

Numerous examples have been found of chemical-dissolution etch pits forming at dislocations under similar conditions of impurity adsorption and small undersaturation. Gilman and Johnston[208] observed such pits on (100)

* The related and quite extensive field of chemical etch pits will not be detailed here. The reader is referred to the recent review by Johnston.[479]

surfaces of lithium fluoride, and Gilman *et al.*[484] showed that adsorption of ferric ion was necessary for this etch-pit formation. Ives and Hirth[299] confirmed the role of ferric ion using radioactive-tracer techniques and showed that etch pits form only at undersaturations less than about 20%. Also we note the very significant work of Dash[485] who demonstrated that chemical etch pits on silicon single crystals corresponded to copper-decorated dislocations as observed by infrared transmission microscopy (see Fig. 45). However, the evidence for correlation with dislocation sites is not as strong in the cases of thermal etch pits as it is in these examples of chemical etch pits.

At larger undersaturations where evaporation is proceeding at a rapid rate the situation is not as clear. Clean high-index surfaces do not show etch pits,[478] as would be expected because of the many ledges present on such a crystal. However, it has been observed in some cases that low-index surfaces exhibit macroscopic etch pits, in spite of the supposed requirement of small undersaturation noted above. Evidently these macroscopic etch pits are not associated with single dislocations. Using a decoration technique Suzuki[471] found as many as ten monatomic spirals contributing to a single macroscopic etch pit on copper. Votava and Amelinckx[466,467] similarly observed that several monatomic spirals contribute to a single macroscopic etch pit. Vassamillet and Hirth[486] found, by the Lambot X-ray technique,[487] dislocation densities of about 10^8 cm^{-2} on (111) and (100) surfaces of silver crystals. On the other hand, the density of macroscopic pits of the type seen on copper by Suzuki[471] was only 10^6 cm^{-2}. Thus macroscopic etch pits, which may form by Frank's[258] impurity-adsorption mechanism, probably do not represent single dislocations.

In summary, single dislocations give rise to etch pits in the presence of an impurity adsorbate when the surface is under a small undersaturation of vapor. The situation is still open to question in the case of macroscopic pits, but present evidence indicates that macroscopic pits are associated with more than one dislocation.

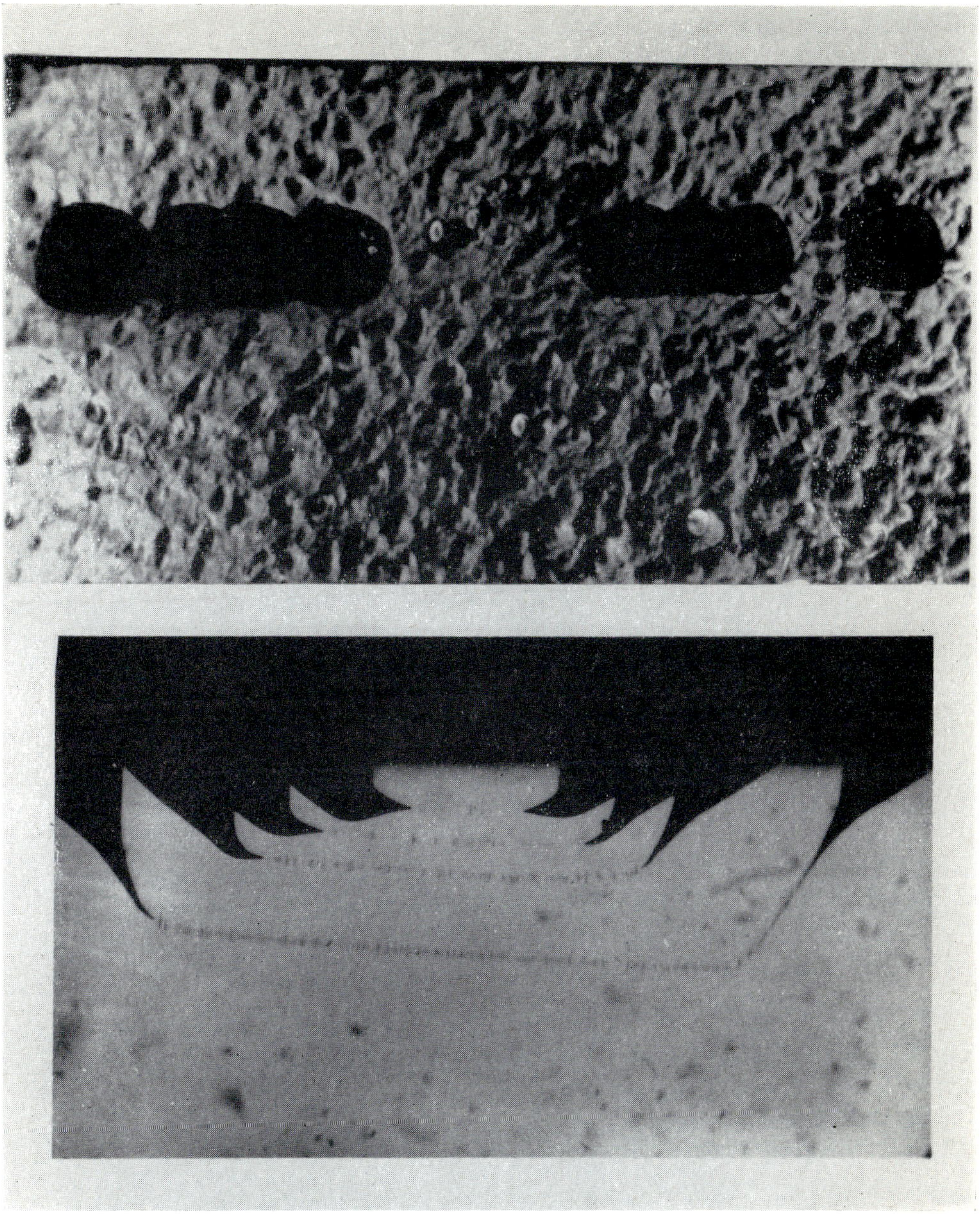

FIG. 45. Chemical etch pits at copper-decorated dislocations observed by infrared transmission microscopy.[485]

F. EBULLITION AND CAVITATION IN LIQUIDS

1. Introduction

Ebullition (boiling) is the homogeneous or heterogeneous nucleation and growth of vapor bubbles in superheated liquids under positive (compressive) hydrostatic stress. Cavitation is the corresponding process in liquids under negative (tensile) hydrostatic stress. Both processes are of very considerable scientific interest, as shall be seen in the following. Further, each process is important in many technological situations. For example the nature of the ebullition is a prime factor in determining the heat transfer coefficients for boiler tubes, and chemical and mechanical engineers are concerned with promotion of the boiling process. Cavitation at propeller tips is an important cause of power loss in ships, and hence marine engineers and architects endeavor to produce designs which will minimize this effect.

The growth of bubbles in ebullition or cavitation should be in accord with the principles outlined in Section D for evaporation of liquids. However, in practice they are highly irreversible processes, and the present authors know of no definitive experimental study of bubble growth in this situation. Therefore the growth aspect of this transformation will not receive further attention here.

There are a number of lucid accounts of nucleation theory for condensation and evaporation processes, and for this reason a fairly terse description of that theory has been given in the present work. However, the literature contains few satisfactory expositions of the somewhat more complicated theory of nucleation in ebullition and cavitation. Accordingly the following description of the theory of homogeneous nucleation in ebullition and cavitation will be more pedagogical in its style.

One notes at the outset that, as in all nucleation theory, the prime assumption is that macroscopic thermodynamic properties may be assigned to the embryos. The principal differences between the models for nucleation in condensation and evaporation on the one hand and ebullition and cavitation on the other arise from the following considerations: (i) In the latter processes the number of molecules in the embryonic vapor bubble is determined not only by the size of the embryo but also by the hydrostatic pressure of the liquid. (ii) As in the cases of condensation and evaporation, the chemical potential of the molecules in the embryo is determined by the curvature of the surface of the condensed phase. However, in ebullition or cavitation the condensed phase is, of course, not the embryo but the supersaturated parent phase. Accordingly, for embryos of size up to the critical, there is *always* equilibration of chemical potential between molecules in the embryo and in the adjacent liquid. This equilibration is conveniently expressed in terms of the vapor pressure in the embryonic bubble

$$p = p_e \exp(-2\sigma\Omega/rkT) \tag{F-1}$$

in which p_e is the equilibrium vapor pressure of bulk liquid *under the imposed hydrostatic stress P*, σ is surface tension, Ω is molecular volume of the liquid and r is radius of curvature of the spherical bubble. This is in sharp contrast to the situation in condensation and evaporation where equilibration of chemical potential of molecules obtains only for the critical nucleus.* (iii) In condensation of liquids there is always mechanical equilibrium between the supersaturated vapor phase and the embryo or nucleus such that the hydrostatic stress in the embryo

$$P_1 = P_2 + 2\sigma/r \tag{F-2}$$

where P_2 is the pressure of the supersaturated phase and r is droplet radius. Similarly in the formation of a disc-shaped evaporation embryo on a close-packed solid surface, equation (F-1) holds for all embryo and nucleus sizes.† On the other hand for nucleation of bubbles in liquids, the mechanical equilibrium implied by equation (F-2) obtains *only* for the critical nucleus. In these cases the appropriate analog of equation (F-2) may be expressed as

$$p^\star = P + 2\sigma/r^\star \tag{F-3}$$

in which $p^\star$ is the partial pressure of vapor in the critical bubble, P is the hydrostatic stress imposed on the liquid and $r^\star$ is the radius of the critical bubble. For $r < r^\star$, $p < P + 2\sigma/r$, and for $r > r^\star$, $p > P + 2\sigma/r$. (iv) For cavitation, where $p_e > P = -|P|$, there is no maximum in the curve of Gibbs free energy of formation versus bubble radius. Rather there is only a critical radius $r^\star$ beyond which $p > P + 2\sigma/r$. At this point the curve ends abruptly because there is no reversible path by which to calculate the net reversible work effect, and hence the Gibbs free energy change, subject to the constraints of constant temperature and the same hydrostatic stress P. Considering the system as liquid, critical bubble, cylinder and piston in an isothermal enclosure of constant volume, work W may be supplied to the piston in a reversible process such that in growth of the bubble to macroscopic size

$$W = \sigma A - E_{el} - ikT \tag{F-4}$$

where A is the final liquid–vapor interfacial area, E_{el} is the elastic strain energy which was in the liquid and i is the number of molecules in the final bubble. One notes that the potential of the system to do net reversible work

* An equivalent situation would exist in condensation if vacancy equilibration obtained within embryos of all sizes, even though molecular equilibrium between embryos and supersaturated vapor occurs only for the embryos of critical size.

† Here P_1 is the pressure of the vapor phase and P_2 is the hydrostatic stress in the solid just beneath the curved peripheral surface of the disc.

may either increase or decrease as the critical bubble grows to macroscopic size. Thus one sees that cavitation of liquids under tensile stresses is a true case of fracture in which the rate is usually determined by the kinetics of formation of supercritical bubbles by statistical fluctuations. (v) Finally, the primitive step in growth of an embryo in cavitation is generally not the transfer of a molecule from the liquid to the vapor phase of the bubble. We shall see that it is the transfer of a vacancy (or part of a vacancy) to the embryonic bubble.

2. Ebullition or Boiling

(a) *Theory of Homogeneous Nucleation in Boiling of Unary Liquids*

The Gibbs free energy of formation of a spherical bubble of radius r in the superheated liquid under positive hydrostatic pressure P is

$$\Delta G^0 = 4\pi r^2\sigma + 4\pi r^3 \Delta G_v/3 \tag{F-5}$$

in which the volume free energy change

$$\Delta G_v = -(kT/\Omega_v) \ln (p_e/P) \tag{F-6}$$

where p_e is the vapor pressure of the superheated liquid. Ω_v is the molecular volume of the vapor in the bubble and to the approximation of the perfect-gas law is represented by

$$\Omega_v = kT/p. \tag{F-7}$$

Here p the pressure of vapor in the bubble is given by equation (F-1). Now ΔG_v is always negative for a superheated liquid and thus equation (F-5) will exhibit a maximum at $r^\star$ where $(\partial \Delta G^0/\partial r) = 0$ and $\Delta G^0 = \Delta G^\star$. Therefore the form of the plot of equation (F-5) is similar to that for nucleation in condensation* (e.g. see equation B-1). However, in this case, rather than applying the usual condition $(\partial \Delta G^0/\partial r) = 0$, the critical values $r^\star$ and $\Delta G^\star$ are more easily obtained by considering an alternative expression for equation (F-5) and applying the condition for mechanical equilibrium (equation F-3) which holds only for the embryonic bubble of critical size. The net reversible work of formation of an embryonic bubble of any size in a system at T and P is

$$\Delta G^0 = 4\pi r^2\sigma + 4\pi r^3(P - p)/3 \tag{F-8}$$

where the last term represents the difference between the work expended against the ambient pressure P and the work gained by allowing the void to fill with vapor at pressure p. Applying equation (F-3) to yield the critical conditions

$$r^\star = 2\sigma/(p^\star - P) \tag{F-9}$$

* Note that, unlike the case of nucleation in condensation, the chemical potential of the critical nucleus is not that of the supersaturated phase.

and

$$\Delta G^\star = 4\pi r^{\star 2}\sigma/3 = 16\pi\sigma^3/3(p^\star - P)^2. \tag{F-10}$$

$p^\star$ may be computed by solution of the transcendental equation

$$(p^\star - P)\Omega = kT \ln (p_e/p^\star), \tag{F-11}$$

which is obtained by combining equations (F-1) and (F-3). The metastable equilibrium number of critical bubbles per cubic centimeter of liquid is

$$n^\star = n_0 \exp(-\Delta G^\star/kT) \tag{F-12}$$

where n_0 is the concentration of monomer in the liquid. This expression ignores any contributions to the free energy of formation of the nucleus from translational and rotational partition functions of the type discussed in the section dealing with homogeneous nucleation of droplets from vapor. It would seem reasonable to neglect these statistical mechanical effects in ordinary viscous liquids.* Also there is no contribution from the free energy of separation (equation B-17) when the free energy of formation of the critical nucleus is calculated in the preceding manner. However, a statistical mechanical term n_0/n_{vac}, similar to that discussed in connection with equation (C-14,), is already included in equation (F-12) to account for the distribution of voids among the n_0 available sites in the liquid.

The frequency of addition of a single molecule to unit area of critical bubbles is

$$\omega = p^\star/(2\pi mkT)^{\frac{1}{2}} = n_s\bar{\omega} \exp(-\Delta G_{\text{vap}}/kT) \tag{F-13}$$

in which n_s is the number of molecules per cm² in the liquid surface, $\bar{\omega}$ is the transmission frequency for the activated species and $\Delta G^\star_{\text{vap}}$ is the free energy of activation for evaporation. To a good approximation

$$\omega = n_s\nu \exp(-\Delta H_{\text{vap}}/kT) \tag{F-14}$$

where ν is the vibrational frequency of molecules in the liquid and ΔH_{vap} is the enthalpy of vaporization. The area of the critical nucleus $A^\star = 4\pi r^{\star 2}$ and the Zeldovich non-equilibrium factor assume the same form as for condensation (equation B-12). Accordingly the rate of homogeneous nucleation of bubbles in boiling becomes

$$\begin{aligned} J &= Z\omega A^\star n^\star \\ &= (\Delta G^\star/3\pi kTi^{\star 2})^{\frac{1}{2}} \cdot p^\star/(2\pi mkT)^{\frac{1}{2}} \cdot 4\pi r^{\star 2} \cdot n_0 \exp(-\Delta G^\star/kT) \end{aligned} \tag{F-15}$$

where $i^\star = 4\pi r^{\star 3}/\Omega_v$ is the number of molecules in the critical bubble. Alternatively in terms of equation (F-14),

$$J = (\Delta G^\star/3\pi kTi^{\star 2})^{\frac{1}{2}} \cdot n_s\nu \exp(-\Delta H_{\text{vap}}/kT) \cdot 4\pi r^{\star 2} \cdot n_0 \exp(-\Delta G^\star/kT). \tag{F-16}$$

* Large negative contributions might be expected in nucleation of bubbles in liquids of zero viscosity such as helium.

Doering[4,488,489] derived a similar relationship

$$J = [6\sigma/\pi m(3 - b)]^{\frac{1}{2}} \exp(-\Delta H_{\text{vap}}/kT) n_0 \cdot \exp(-\Delta G^\star/kT) \quad \text{(F-17)}$$

in which $b = (p^\star - P)/p^\star$. However, the authors are of the opinion that the origin of $\exp(-\Delta H_{\text{vap}}/kT)$ is not clear via the Doering–Volmer derivation.* Also it is noted that equation (F-17) does not hold for the possible situation that $b = 3$; this arises from an approximation in the derivation which may be avoided through a graphical method discussed in the original reference[4] and in reference 490. Thus the present authors feel that equation (F-16), which was derived with only the usual assumption that macroscopic thermodynamic properties may be ascribed to embryos, is superior to equation (F-17) and should be preferred for description of the kinetics of homogeneous nucleation in the process of ebullition. As in condensation, however, it should be borne in mind that the assumption that macroscopic thermodynamic properties can be ascribed to the embryos may not hold; in such a case the σ value obtained from a correlation of equation (F-16) with experiment should be regarded as a phenomenological parameter representing the barrier to nucleation.

Nevertheless, the macroscopic approximation is usually excellent for ebullition where the critical nucleus may be of the order of a micron in diameter and contain millions of molecules. Also, unlike the case of nucleation in the vapor, there should be no difficulty relating to thermal non-accommodation because the embryonic bubbles are in good thermal contact with the condensed phase. Finally, the period of most ebullition experiments is sufficiently long that there should be no problem having to do with transients. Therefore it would seem that ebullition studies offer a promising method for the further establishment of nucleation theory.

(b) *Experiment in Boiling of Unary Liquids*

The experimental data which are thought to relate to homogeneous nucleation in boiling of liquids are quite sparse. Perhaps the best known data are those of Wismer[491] and co-workers[492] on the critical superheating of ethyl ether. These workers used a U-shaped capillary tube which was connected to a screw compressor as a container for the liquid. The experiments were conducted by imposing an initial high pressure on the liquid in the tube, which was held at a constant temperature in a thermostat bath, and then suddenly reducing the pressure to a low value by means of the compression screw. Higher pressures were measured by a metal pressure gauge and lower pressures by a mercury manometer. The terminal pressure that was critical for spontaneous formation of bubbles was recorded. This datum, together with the temperature, described the critical superheating for nucleation of

* Substituting in equation (F-16) for $i^\star$, $r^\star$ and $\Delta G^\star$, one finds that the ratio of J as given by (F-16) to J as given by (F-17) is $[(3 - b)/27]^{\frac{1}{2}} \exp(\Delta H_{\text{vap}}/kT)$.

bubbles. The scatter of the data was appreciable but in view of the fact that most errors would lead to low values of superheating only the upper limits (lowest pressures) were taken as significant with respect to homogeneous nucleation. The data were compared[4] with equation (F-17) (or equation F-16) by setting $J = 10\ \text{cc}^{-1}\ \text{sec}^{-1}$. The results are given in Fig. 46 where

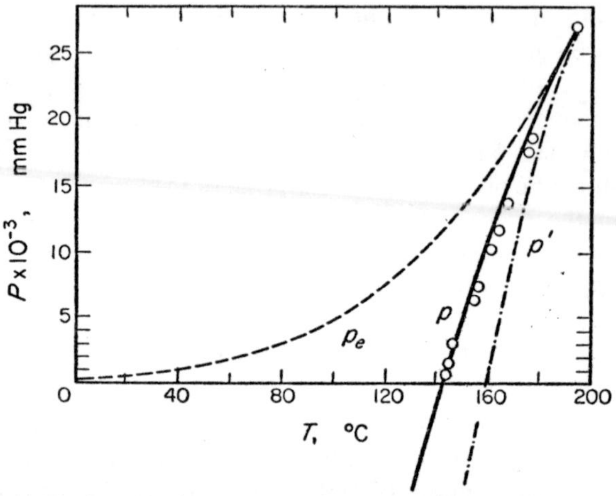

FIG. 46. Comparison of homogeneous nucleation theory with observations on the critical superheating of ethyl ether.

calculated values are represented by the solid lines and the data by circles. It is seen that the agreement is very good.* Here p_e is the equilibrium vapor pressure of ethyl ether, P is the terminal hydrostatic pressure which is critical for appreciable rate of bubble formation and P' is the minimum pressure for bubble formation as calculated from the van der Waals equation. It is evident that vaporization occurs by nucleation and growth of bubbles in accordance with equation (F-3) and hence before the vapor pressure of the liquid equals the sum of hydrostatic and "internal" pressures. One notes that Temperley[493] interpreted these data as the temperatures at which the minimum of the pressure on the van der Waals curve is practically zero. The present authors are inclined to view this limited agreement as fortuitous.

(c) *Heterogeneous Nucleation at Solid–Liquid Interfaces*

It remains to consider the role of heterogeneous nucleation in causing the scatter in the data of Wismer *et al.* Frenkel[72] and Fisher[494] have considered the effect of solid–liquid interfaces in heterogeneous nucleation of bubbles in liquids. Analogously to the development of equation (C-6) for the

* These data should be corrected to consider the temperature drop in the liquid on expansion (~1°); this correction would enhance the agreement.

free energy of formation of cap-shaped nuclei of crystals on flat substrate surfaces it may be shown that the free energy of formation of a critical bubble at a planar solid–liquid interface is given by

$$\Delta G^\star = 16\pi\sigma^3\phi_3(\theta')/3\Delta G_v^2 \tag{F-18}$$

or

$$\Delta G^\star = 16\pi\sigma^3\phi_3(\theta')/3(p^\star - P)^2 \tag{F-19}$$

where

$$\phi_3(\theta') = (2 + \cos\theta')(1 - \cos\theta')^2/4 \tag{F-20}$$

and θ' is π minus the equilibrium contact angle θ between the liquid and the solid. Thus in the case of complete wetting $\theta' = 180°$ and $\phi_3(\theta')$ is unity and the solid–liquid interface has no catalytic potency for nucleation of bubbles. If there is no wetting $\phi' = 0°$ and $\phi_3(\theta')$ is zero and the solid–liquid interface is a perfect catalyst for nucleation. In intermediate cases $\phi_3(\theta')$ has the same dependence as that shown in Fig. 11 for $\phi_3(\theta)$. Similarly to the argument leading to equation (C-11) or (C-19) it may be shown that the nucleation rate of bubbles at a planar solid–liquid interface is

$$J = (\Delta G^\star/3\pi kTi^{\star 2})^{\frac{1}{2}} \cdot p^\star/(2\pi mkT)^{\frac{1}{2}} \cdot 2\pi r^{\star 2}(1 - \cos\theta') \cdot n_s \exp(-\Delta G^\star/kT). \tag{F-21}$$

There are no quantitative data to support this equation but it is evident that solid–liquid interfaces (usually occasioned by the presence of impurity particles) may readily catalyze nucleation of bubbles and give rise to scatter in critical superheating data.

Actually, the critical superheatings observed in the presence of the usual rough solid–liquid interfaces should be much less than those predicted by

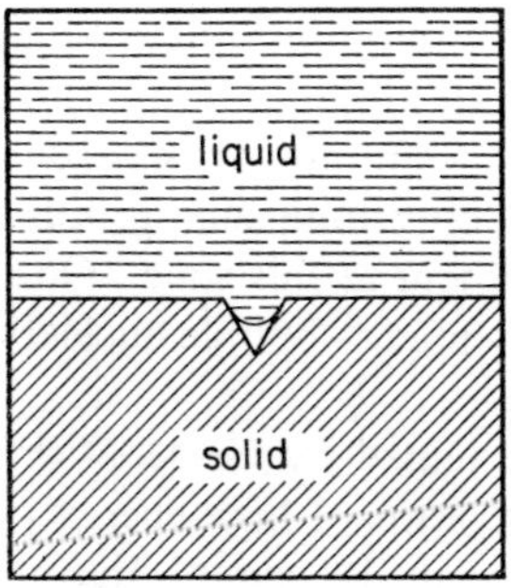

FIG. 47. Conical cavity in a solid–liquid interface.

equation (F-21) because, analogously to the case of condensation discussed in Section C-2-a-4, re-entrant angles and other surface imperfections should be more potent nucleation catalysts than a flat substrate. Fisher[494] has demonstrated this principle for the case of a conical cavity (Fig. 47) in a solid–liquid interface for which the equilibrium contact angle θ is greater than 90°. It is seen that huge pressures would be required to force the liquid to the bottom of the cavity. In the absence of such a pressure a vapor pocket would remain in the cavity and if the vapor pressure of the liquid should be made to exceed the hydrostatic stress in the liquid the vapor pocket would simply grow to form a macroscopic bubble.

Lewis(495) has studied the nucleation of voids in water under tension in Bertholet tubes(496) at 27 to 34°C and has observed that the efficacy of Pyrex and steel tubing as a nucleation catalyst is sensitive to the state of imperfection of the surfaces.

(d) *Ebullition of Dissolved Gases*

The interesting case of ebullition of a gas dissolved in a liquid phase may also be treated by equations (F-16) and (F-21) with the change that p_e in equation (F-6) should now correspond to the partial pressure of the gas in equilibrium with the supersaturated bulk liquid *under the imposed hydrostatic stress P*. In addition, σ in equation (F-10) will no longer be a constant but will vary with the partial pressure of gas in the embryo in accordance with an adsorption isotherm such as that of Gibbs. Of course an intermediate case could obtain in which both the solvent and the solute have appreciable partial pressures in the critical embryo. The authors are aware of no quantitative study of this problem, but numerous qualitative observations of catalysis of gas ebullition by a substrate have been made. Examples are the nucleation of CO_2 bubbles in a glass of beer or the nucleation of CO bubbles in the deoxidation of steel in an open-hearth furnace.

(e) *Nucleation in Bubble Chambers*

Again analogously to the development in Section B, given the basic formalism leading to equations (F-16) and (F-21) one can extend the treatment to heterogeneous nucleation on tiny impurity particles, on ions, etc. However, the limited data available concerning such processes do not seem to warrant a detailed theoretical treatment at this time. One interesting example, however, which has been represented as bubble nucleation on ions, is the hydrogen bubble chamber.

The bubble chamber, invented by Glaser(497,498) in 1952, has become an important tool in the observation of high-energy particles. In this technique a volatile liquid such as propane is suddenly decompressed to the superheated condition and passage of high-energy particles through the liquid nucleates bubbles which are observable as "tracks". Thus the principle of operation is somewhat similar to that of a cloud chamber. However, as will be discussed in the following, it seems likely that the nucleation is not on ions but in thermal spikes in the liquid. Perhaps the most complete investigations of nucleation in bubble chambers are those by Alfredsson and Johansson(499) and Johansson.(500) These workers measured the bubble density due to 18 MeV electrons as a function of temperature and pressure drop in liquid propane, Freon-12 and sulfur hexafluoride. Although their theoretical interpretation of the data is rather qualitative, they considered the possibility of bubble nucleation by either ions or thermal spikes.

Assuming that n elementary charges are uniformly distributed over the bubble surface, the analog of equation (F-3) for the pressure inside a critical nucleus is

$$p^{\star} = P + 2\sigma/r^{\star} - n^2(1 - \mathscr{E}/n)e^2/8\pi K r^4 \qquad \text{(F-22)}$$

where $\mathscr{E}$ is the dielectric constant of the liquid. Upon maximizing (F-22) and assuming isothermal conditions the limiting degree of superheat at which a bubble will surely grow to visible size is found in terms of pressure to be

$$p^{\star} - P = (27\pi/2e^2)^{\frac{1}{3}}\sigma^{\frac{4}{3}}\mathscr{E}^{\frac{1}{3}}[n^2(1 - \mathscr{E}/n)]^{-\frac{1}{3}}. \qquad \text{(F-23)}$$

It may be shown from (F-23) that the decompression $\Delta P = p_e - P$ which is required for bubble formation is given as a function of temperature by

$$\log \Delta P = \text{constant} + K_1 \log (T_c - T) \qquad \text{(F-24)}$$

in which K_1 is a constant characteristic for nucleation by ions and T_c is the critical temperature of the liquid.

Alternatively the process may be viewed in terms of bubble nucleation in thermal spikes. Johansson[500] considered that the high-energy particle supplies the energy for formation of the bubbles in an "adiabatic" manner, i.e. none of the energy is provided by thermal fluctuations in the liquid. Summing the surface energy, the energy required to vaporize the gas in the bubble and the work done against the hydrostatic pressure, the total energy of formation is

$$W = (32\pi\sigma^3/3\Delta P^2)[(\Delta H_{\text{vap}}\rho_{\text{vap}}/\Delta P) + \tfrac{1}{2} - (3T/2\sigma)(\mathrm{d}\sigma/\mathrm{d}T)] \qquad \text{(F-25)}$$

in which ΔH_{vap} is the heat of vaporization of the liquid and ρ_{vap} is the density of the vapor. Equation (F-25) may be solved to yield the temperature dependence of the decompression required for bubble formation. This relationship is identical in form with equation (F-24) except that the constants have different values. For this case K_1 is replaced by K_2 which has a value characteristic for nucleation in thermal spikes.

The experimental results[499] are better described by K_2, indicating that nucleation of bubbles occurred in thermal spikes and that ions were not involved. However, one notes that kinetic factors have been generally ignored in both treatments. As in the case of cloud-chamber studies, perhaps the problem should be viewed as a transient.

3. Nucleation in Cavitation of Unary Liquids

(a) *Theory*

Despite the peculiarities of this case that were pointed out in the Introduction to the present section, the derivations and resulting equations relating to nucleation of bubbles in liquids under hydrostatic tension are almost identical with those considered for the case of boiling. Further, even the qualitative considerations having to do with catalysis of bubble formation at rough solid–liquid interfaces are the same as in the case of boiling. However, there are two important differences which arise from the consideration that the primitive step in growth of an embryonic bubble in cavitation is generally transport of a vacancy (or part of a vacancy) rather than transport of a molecule to the vapor phase. That this is so may be seen by considering that the critical nucleus in ebullition contains many molecules (of the order of 10^6 in some cases) while the critical nucleus in cavitation may contain on the average of the order of one or less. Accordingly the frequency of occurrence of this primitive step per unit area of free embryo surface is

$$\omega = n_s \nu \exp(-\Delta G_d^\star/kT) \tag{F-26}$$

in which $\Delta G_d^\star$ is the free energy of activation for diffusion in the liquid and is usually of the order of 5 kcal per mole. Further, the non-equilibrium factor must be modified to give

$$Z = (\Delta G^\star/3\pi kT i_{\text{vac}}^{\star 2})^{\frac{1}{2}} \tag{F-27}$$

where $i_{\text{vac}}^\star$ is the number of vacancies (or primitive units of volume increase) in the critical nucleus. Thus the analog of equation (F-16) for homogeneous nucleation becomes

$$J = (\Delta G^\star/3\pi kT i_{\text{vac}}^{\star 2})^{\frac{1}{2}} \cdot n_s \nu \exp(-\Delta G_d^\star/kT) \cdot 4\pi r^{\star 2} \cdot n_0 \exp(-\Delta G^\star/kT). \tag{F-28}*$$

Similarly the analog of equation (F-21) for heterogeneous nucleation at planar solid–liquid interfaces is

$$\begin{aligned} J = (\Delta G^\star/3\pi kT i_{\text{vac}}^{\star 2})^{\frac{1}{2}} &\cdot n_s \nu \exp(-\Delta G_d^\star/kT) \\ &\cdot 2\pi r^{\star 2}(1 - \cos\theta') \\ &\cdot n_s \exp(-\Delta G^\star/kT). \end{aligned} \tag{F-29}$$

Equations (F-28) and (F-29) were first derived by Fisher[(494)] in a rather simplified form under the reasonable assumptions that Z and $n_s A^\star$ may be taken as unity and $\nu \simeq kT/h$. His expression for homogeneous nucleation is

$$J = (NkT/h) \exp[-(\Delta G_d^\star + \Delta G^\star)/kT] \tag{F-30}$$

where N is Avogadro's number and hence J has the dimensions of reciprocal mole-seconds. The corresponding equation for critical fracture pressure at which $J = 1$ is

$$P \simeq -[16\pi\sigma^3/3kT \ln(\text{B})]^{\frac{1}{2}} \tag{F-31}$$

* The concentration of monomer rather than vacancies appears in this equation due to a statistical factor n_0/n_{vac} similar to that discussed in connection with equations (C-14) and (F-12).

where, using Fisher's equation (F-30) and neglecting $\Delta G_d^\star$, $B = (NkT/h)$, or using equation (F-28), $B = (\Delta G^\star/3\pi kTi_{\text{vac}}^{\star 2})^{\frac{1}{2}} n_s \nu \exp(-\Delta G_d^\star/kT) 4\pi r^{\star 2} n_0$.

One notes that as in ebullition thermal non-accommodation and transients should not be problems here. Again, there are no statistical mechanical contributions other than a statistical factor analogous to that of equation (C-14). However, unlike the situation in ebullition the critical bubble in cavitation may be of the order 10^{-7} cm in radius and contain less than a molecule. Accordingly the macroscopic approximation is really questionable in this case.

Before proceeding to a consideration of the fracture-pressure data it seems necessary to mention somewhat similar equations which were derived by Doering.[4,488,489,490] His method resembled those outlined above but he assumed that the primitive step in embryo growth was transport of a molecule to the vapor phase of the embryonic bubble. As discussed in the foregoing, the present authors consider this assumption to be erroneous for the usual experimental cases. Nevertheless, his expression for critical fracture pressure gives numerical answers which are quite similar to those obtained from equation (F-31). This is reasonable in view of the insensitivity of the fracture pressure to the pre-exponential terms in the expressions for J. One notes that Wakeshima[501] has revised the Fisher–Doering[488,490] calculation by considering the effect of curvature on surface tension through a modification of Tolman's quasi-thermodynamic treatment.[76] As discussed in connection with the theory of homogeneous nucleation of droplets from unary vapors, it is doubtful whether such thermodynamic methods of correction of surface tension provide any significant improvement of the existing descriptions.

(b) *Experiment*

At least four techniques may be used to measure the tensile strength of liquids. (i) In the older literature a standard method was to slowly cool glass tubes which had been filled with liquid and note the temperature at which rupture occurred. The fracture pressure was then computed from the coefficients of thermal expansion and compressibility of glass and liquid.[502] (ii) A later method was to fill a metal bellows with liquid and measure the tensile strength directly.[503,504] Neither technique gave very reproducible results and more importantly the observed fracture pressures were generally lower than those obtained by newer methods, indicating that bubble nucleation probably occurred heterogeneously at potent catalytic sites. (iii) A most promising method, which has received insufficient attention, is the observation of absorption peaks associated with high-frequency acoustical waves. (iv) Perhaps the most reliable of the fracture pressure data have been obtained by Briggs[505–509] who used a centrifugal technique. Tubes filled with liquid were rotated in a centrifuge at increasing speeds until rupture of the liquid occurred. A summary of the fracture pressures for various liquids at 20°C is given in Table 13

TABLE 13

FRACTURE PRESSURES OF VARIOUS LIQUIDS IN ATMOSPHERES AT 20°C

Theoretical values from Fisher's equation (F-31) for homogeneous nucleation. Data from Briggs.

Liquid	Theoretical fracture pressure, atm	Observed fracture pressure, atm
Water	−1,380	−270
Chloroform	−318	−290
Benzene	−352	−150
Acetic acid	−325	−288
Aniline	−625	−280
Carbon tetrachloride	−315	−275
Mercury	−23,100	−425

together with the theoretical values computed from Fisher's equation (F-31) for homogeneous nucleation. It is seen that the observed tensile strengths are appreciably lower than the theoretical in all cases. One concludes that the nucleation of bubbles probably occurred heterogeneously at solid–liquid interfaces.

4. SUMMARY OF EBULLITION AND CAVITATION

There are no new statistical mechanical corrections for nucleation in boiling or cavitation but only the statistical factor, which had been included previously, as mentioned in connection with equations (F-12) and (F-28).

Equation (F-15) or (F-16) is recommended for homogeneous nucleation rate in boiling of superheated liquids. The size of the critical nucleus (of the order of a micron in diameter) is sufficiently large that macroscopic thermodynamic quantities may be employed in its description with some degree of confidence. Also, due to the nature of the critical superheating experiments, it would seem that thermal accommodation and transient phenomena are not a problem. Accordingly critical superheating experiments may provide the most unambiguous test of homogeneous nucleation theory. It is noteworthy that these equations describe the data of Wismer[491,492] *et al.* quite well.

Nucleation in bubble chambers has been studied experimentally and interpreted theoretically in an approximate fashion by Alfredsson[499] and Johansson.[499,500] They conclude that the bubbles nucleate in thermal spikes rather than about ions. A kinetic treatment of this problem as a transient seems definitely in order.

Equation (F-28) is recommended for homogeneous nucleation rate of bubbles in liquids under tensile hydrostatic stress and equation (F-30) predicts the fracture pressure. Although there should be no problems related to non-accommodation and transients in the fracture-pressure experiments, the size of the critical nucleus is generally so small that application of macroscopic thermodynamic concepts is again questionable. Also, the primitive unit of growth of the bubble embryo is probably some unknown fraction of a vacancy.

The fracture pressures observed by Briggs[509] are all lower than the theoretical, and one is inclined to interpret this as due to heterogeneous nucleation of bubbles at solid–liquid interfaces.

G. VOID FORMATION IN SOLIDS

1. Introduction

The final phase transformation involving the vapor that shall be considered in this monograph is the process of void information in solids. A special instance of this type, void formation at a free surface, has already been discussed in Section D-3-b-2. The growth aspect of this transition is thought to be simply a boundary value problem involving the diffusion of vacancies and hence will not be treated further. Also, data on the nucleation kinetics involved in void formation is sparse, and therefore only a brief account of these processes will be undertaken here. Void formation has been observed in diffusion couples[510,511] and in creep[512] at elevated temperatures where, because the specific interfacial free energy is more isotropic at higher temperatures, the voids tend to form in the shape of segments of spheres. Also, vacancies quenched-in from elevated temperatures have been observed to lead to the formation of dislocation loops within monocrystalline grains.[513,514] According to a mechanism first proposed by Seitz[515] and Frank[516] (see also refs. 517, 518, 519) these dislocation loops form by the nucleation of vacancy discs,* which subsequently collapse to form the loops.

2. Nucleation of Voids at Elevated Temperature

At elevated temperatures where vacancy mobility as determined from the diffusion coefficient is high, metastable equilibrium between single vacancies and embryos may be assumed. The nucleation kinetics are then given by equation (F-28) for homogeneous nucleation and (F-29) for heterogeneous nucleation except that in this case

$$\Delta G_v = (-kT/\Omega) \ln (n_{\text{vac}}/n_{\text{vac}_e}) \qquad \text{(G-1)}$$

where n_{vac} and n_{vac_e} represent, respectively, the actual and equilibrium concentration of vacancies.† An equation similar to (F-29) but differing in the pre-exponential factor has been derived by Resnick and Seigle.[520]

Resnick and Seigle studied pore formation during the interdiffusion of copper and zinc. They found a critical vacancy supersaturation ratio $(n_{\text{vac}}/n_{\text{vac}_e})_{\text{crit}}$ of about 1.5 and a contact angle $\theta = 150°$ for the nucleation of

* Bounded by low-index planes.

† Note that $n_s\nu \exp(-\Delta G_d^\star/kT)$ (equation F-28), the frequency with which an atom at the void surface diffuses into the crystal, equals $n_{\text{vac}}\nu \exp(-\Delta G_{\text{vd}}^\star/kT)$ in which $\Delta G_{\text{vd}}^\star$ is the activational free energy for vacancy diffusion.

pores on a heterogeneity which was presumed to be zinc oxide. However, their correlation was only approximate, and thus it appears that further work is required for a definitive test of equation (F-29) as modified for this case.

3. Nucleation of Vacancy Discs

The rate equation for vacancy-disc nucleation should differ from equation (F-28) only in the geometry of the critical nucleus. The free energy of formation of the critical disc is

$$\Delta G^\star = 2\pi r^{\star 2}\sigma + 2\pi r^\star \varepsilon + \pi r^{\star 2} h \Delta G_v = -\pi\varepsilon^2/(2\sigma + h\Delta G_v) \quad \text{(G-2)}$$

where ε is the edge energy of the disc, σ is the interfacial free energy of the low-index circular surface and h is the height of the disc. This expression is valid only when $|h\Delta G_v| > |2\sigma|$.

Although numerous experiments have been performed in which vacancies have been quenched-in from elevated temperatures (see ref. 521), the vacancies may also diffuse to sinks such as the free surface, dislocations and grain boundaries. This consideration enormously complicates the analysis of vacancy removal.(521,522) Thus even though vacancy-disc dislocation loops have been observed,(513,514) the kinetics of disc nucleation have not been quantitatively studied.

Nevertheless data have been obtained(523) on the quenching-in of vacancies which may pertain to the interesting postulation of athermal nucleation by Fisher, Hollomon and Turnbull.(99) They developed the kinetics of athermal nucleation for precipitation in condensed phases but their treatment can easily be extended to vacancy-disc nucleation. Let a crystal be equilibrated at temperature T_Q where the equilibrium concentration of vacancies is n_{vac} and then quenched to a temperature T. The metastable-equilibrium concentration of critical nuclei at T is

$$n^\star = n_0 \exp(-\Delta G^\star/kT) \quad \text{(G-3)}$$

where n_0 is the concentration of lattice sites and $\Delta G^\star$ is given by equation (G-2). Now at the temperature T_Q the concentration of embryos (subcritical at T_Q of course) of the *same size* is

$$n_i = n_0 \exp(-\Delta G^0/kT_Q). \quad \text{(G-4)}$$

If the quench from T_Q to T is very rapid so that embryos existing at T_Q do not dissociate during quenching, embryos which are critical at T will already exist in appreciable number if n_i (equation G-4) is greater than $n^\star$ (equation G-3). Since the nuclei n_i do not require thermal fluctuation to form at T, this process is called athermal nucleation.(99) The condition that $n_i > n^\star$ leads to the requirement that

$$(\Delta G^0/T_Q) < (\Delta G^\star/T) \quad \text{(G-5)}$$

for athermal nucleation. Now ΔG^0 is given by equation (G-2) which becomes

$$\Delta G^0 = 2\pi r^{\star 2}\sigma + 2\pi r^{\star}\varepsilon \quad \text{(G-6)}$$

in this case because $\Delta G_v = 0$ at T_Q. Also

$$r^{\star} = -\varepsilon/(2\sigma + h\Delta G_v) \quad \text{(G-7)}$$

and hence upon substituting (G-2) and (G-6) into (G-5) one obtains the conditions for appreciable athermal nucleation

$$(T_Q/T) > 2 - [2\sigma/(2\sigma + h\Delta G_v)]$$

or

$$(T_Q/T) \gtrsim 2 - (2\sigma/h\Delta G_v). \quad \text{(G-8)}$$

The term $(2\sigma/h\Delta G_v)$ is negative and varies in magnitude from a large number when $T \simeq T_Q$ to a value of about unity when $T \ll T_Q$.

Baurle and Koehler[523] performed careful experiments in which they quenched gold wires to room temperature and studied the kinetics of the decrease in vacancy concentration. They found anomalous behavior which indicated the presence of vacancy clusters in the as-quenched wires for $T_Q \gtrsim 700°C$. They and Seitz[524] analyzed this anomaly on the supposition that the clusters formed during quenching. However, noting that vacancy clusters appeared when $(T_Q/T) \gtrsim 3.3$ it would appear in view of equation (G-8) that athermal nucleation, as proposed by Fisher *et al.*,[99] could account for these results.

4. Summary of Void Formation in Solids

Void nucleation and growth in solids should follow essentially the same kinetics as in the liquid. Again, there are no new statistical mechanical contributions to the free energy of formation of the critical nuclei. The kinetics of void nucleation in solids have not been studied quantitatively. However, this appears to be a promising area for study because it is one of the few situations in nature where homogeneous nucleation may occur. Finally, the possibility exists that athermal nucleation of vacancy clusters may occur during quenching experiments.

NOTE ADDED IN PROOF

As this monograph went to press, several articles appeared in print which are of considerable importance.

In Section C we discussed the tenuous assumption that macroscopic thermodynamic properties may be ascribed to critical nuclei containing less than 20 molecules ($i^* < 20$). Also it was noted that no nucleation barrier exists at all and that every incident atom accretes to the substrate when $i^* < 1$. Further, one statistical mechanical model for description of heterogeneous nucleation in the range $1 < i^* < 20$ was cited.(87) Recently, Walton (531, 532) and Rhodin and Walton(533) also considered the extension of nucleation theory into the range $1 < i^* < 20$. They approached the problem via the petit canonical ensemble and treated the embryos as adsorbed macromolecules in terms of potential energies and partition functions. Their result is formally identical with equations (C-6) and (C-16). Such descriptions would, of course, be preferred for such small clusters. However, in both of these statistical mechanical treatments,(87, 531–533) we note the extreme difficulty in even a qualitative evaluation of the internal partition functions and potential energies.(534) Also, we feel that low supersaturations and high substrate temperatures, which result in large values of i^*, are requisite for epitaxy, as discussed in Section C. Thus we are inclined to question Walton's and Rhodin's model(531–533) for epitaxial deposition. This disagreement suggests the need for experimental resolution. Nevertheless, the treatment of Rhodin and Walton(533) should prove very useful in describing the discontinuous changes in properties of the nucleus at small sizes.

Secondly, we cite the descriptions of recent theories and experiments dealing with condensation and evaporation published or to be published this year as the Proceedings of various conferences.(535–538)

Finally, it is noted that Hruska,(539) in experiments in which the critical supersaturation for cadmium nucleation on copper and glass was measured as a function of the beam temperature, has shown that beam temperature does not affect the nucleation kinetics. Hence, as was already suggested in Section C, thermal non-accommodation evidently does not occur in such nucleation.

REFERENCES

1. E. H. Kennard, *Kinetic Theory of Gases*, McGraw-Hill, New York (1938).
2. W. Kossel, *Nach. Ges. Wiss. Göttingen*, 135 (1927).
3. I. N. Stranski, *Z. phys. Chem.* **136**, 259 (1928); **11**, 421 (1931).
4. M. Volmer, *Kinetik der Phasenbildung*, Steinkopff, Dresden and Leipzig (1939).
5. L. Onsager, *Phys. Rev.* **37**, 405 (1931).
6. S. Glasstone, K. J. Laidler and H. Eyring, *Theory of Rate Processes*, McGraw-Hill, New York (1941).
7. H. Hertz, *Ann. Phys.* **17**, 177 (1882).
8. M. Knudsen, *Ann. Phys.* **29**, 179 (1909).
9. M. Knudsen, *Ann. Phys.* **34**, 593 (1911).
10. M. Knudsen, *Ann. Phys.* **48**, 1113 (1915).
11. H. S. W. Massey and E. H. S. Burhop, *Electronic and Ionic Impact Phenomena*, Chapter 9, Oxford Univ. Press, London (1952).
12. F. M. Devienne, *Mem. des. sciences phys., Acad. Sci. Paris* **56**, 1 (1953).
13. K. Schäfer, *Fortschr. chem. Forsch.* **1**, 61 (1949).
14. K. F. Herzfeld, *Eucken-Wolf Hand. und Jahrbuch der Chem. Phys.* **3**, 95 (1937).
15. G. Ehrlich, *Structure and Properties of Thin Films*, edited by C. A. Neugebauer, J. B. Newkirk and D. A. Vermilyea, p, 423, John Wiley, New York (1959).
16. G. W. Sears, *J. Chem. Phys.* **25**, 637 (1951).
17. W. Gaede, *Ann. Phys.* **41**, 331 (1913).
18. M. Knudsen, *Ann. Phys.* **52**, 105 (1917).
19. R. A. Millikan, *Phys. Rev.* **22**, 1 (1923).
20. P. S. Epstein, *Phys. Rev.* **23**, 710 (1924).
21. R. W. Wood, *Phil. Mag.* **30**, 300 (1915); **32**, 314 (1916).
22. J. P. Taylor, *Phys. Rev.* **35**, 375 (1930).
23. G. A. Bassett, J. W. Menter and D. W. Pashley, ref. 15, p. 33.
24. I. Estermann and O. Stern, *Z. Phys.* **61**, 95 (1930).
25. I. Estermann, R. Frisch and O. Stern, *Z. Phys.* **73**, 348 (1931).
26. F. Knauer and O. Stern, *Z. Phys.* **53**, 799 (1929).
27. B. Josephy, *Z. Phys.* **80**, 755 (1933).
28. R. R. Hancox, *Phys. Rev.* **42**, 864 (1932).
29. A. Ellett and H. F. Olson, *Phys. Rev.* **31**, 643 (1928).
30. J. M. B. Kellog, *Phys. Rev.* **41**, 635 (1932).
31. H. A. Zahl, *Phys. Rev.* **36**, 893 (1930).
32. A. Ellett, H. F. Olson and H. A. Zahl, *Phys. Rev.* **34**, 493 (1929).
33. F. C. Hurlbut, *Recent Research in Molecular Beams*, edited by I. Estermann, p. 145, Academic Press, New York (1959).
34. G. K. Wehner, *Phys. Rev.* **102**, 690 (1956).
35. J. E. Lennard-Jones and A. F. Devonshire, *Proc. Roy. Soc. (London)* **A156**, 6 (1936); A. F. Devonshire, *ibid.*, **A158**, 269 (1937).
36. C. Zener, *Phys. Rev.* **40**, 335 (1932).
37. I. Langmuir, *Phys. Rev.* **8**, 149 (1916).
38. N. Cabrera, *Disc. Faraday Soc.* **28**, 16 (1959).
39. R. W. Zwanzig, *J. Chem. Phys.* **32**, 1173 (1960).
40. E. Schrödinger, *Ann. Phys.* **44**, 916 (1914).
41. J. P. Hirth and G. M. Pound, *J. Phys. Chem.* **64**, 619 (1960).
42. S. Miyamoto, *Trans. Faraday Soc.* **29**, 794 (1933).
43. N. Fuchs, *Phys. Z. Sowjet* **4**, 481 (1933).
44. J. E. Lennard-Jones, *Proc. Roy. Soc. (London)* **A163**, 127 (1937).
45. J. H. McFee, Ph.D. Thesis, Carnegie Institute of Technology, 1960, to be published.
46. J. E. Lennard-Jones and C. Strachan, *Proc. Roy. Soc. (London)* **A150**, 442 (1935).
47. G. W. Sears and J. W. Cahn, *J. Chem. Phys.* **33**, 494 (1960).

48. N. F. Ramsey, *Molecular Beams*, p. 35, Oxford Univ. Press, London (1956).
49. O. Stern, *Z. Phys.* **2**, 49 (1920); **3**, 417 (1920).
50. G. M. Rothberg, M. Eisenstadt and P. Kusch, *J. Chem. Phys.* **30**, 517 (1959).
51. T. Alty and C. A. Mackay, *Proc. Roy. Soc.* (*London*) **149A**, 104 (1935).
52. A. Ellett and V. W. Cohen, *Phys. Rev.* **52**, 509 (1937).
53. F. L. Hughes, *Phys. Rev.* **113**, 1036 (1959).
54. H. Fraunfelder, E. Lüscher, E. von Goeler and R. N. Peacock, to be published.
55. R. Smoluchowski, in *Phase Transformations in Solids*, edited by R. Smoluchowski, J. E. Meyer and W. A. Weyl, pp. 149–183, John Wiley, New York (1951).
56. D. Turnbull and J. H. Hollomon, in *The Physics of Powder Metallurgy*, edited by W. E. Kingston, pp. 109–143, McGraw-Hill, New York (1951).
57. R. S. Bradley, *Quarterly Reviews* (*London*) **5**, 315 (1951).
58. V. K. LaMer, *Ind. Eng. Chem.* **44**, 1270 (1952).
59. G. M. Pound, *Ind. Eng. Chem.* **44**, 1278 (1952).
60. J. H. Hollomon and D. Turnbull, in *Progress in Metal Physics*, Vol. 4, edited by B. Chalmers and R. King, pp. 333–388, Pergamon Press, London (1953).
61. W. J. Dunning, in *Chemistry of the Solid State*, edited by W. E. Garner, pp. 159–183, Academic Press, New York (1955).
62. D. Turnbull, in *Solid State Physics*, Vol. 3, edited by F. Seitz and D. Turnbull, pp. 225–306, Academic Press, New York (1956).
63. G. M. Pound, in *Liquid Metals and Solidification*, edited by B. Chalmers, pp. 87–105, American Society for Metals, Cleveland (1958).
64. W. G. Courtney, Texaco Experiment Inc. Report TM-1250, July (1961).
65. J. Lothe and G. M. Pound, *J. Chem. Phys.* **36**, 2080 (1962).
66. M. Volmer and A. Weber, *Z. phys. Chem.* **119**, 277 (1925).
67. R. Kaischew and I. N. Stranski, *Z. phys. Chem.* **26B**, 317 (1934).
68. J. W. Gibbs, *Collected Works*, Vol. 1, *Thermodynamics*, Yale University Press, New Haven (1948).
69. B. E. Sundquist and R. A. Oriani, *J. Chem. Phys.* **36**, 2604 (1962).
70. R. Becker and W. Döring, *Ann. Phys.* (5) **24**, 719 (1935).
71. J. Zeldovich, *J. Exp. Theor. Phys.* (Russ.) **12**, 525 (1942).
72. J. Frenkel, *Kinetic Theory of Liquids*, Oxford Univ. Press, London (1946).
73. H. Reiss, *Ind. Eng. Chem.* **44**, 1286 (1952).
74. A. J. Barnard, *Proc. Roy. Soc.* (*London*) **A220**, 132 (1953).
75. C. N. Yang and T. D. Lee, *Phys. Rev.* **87**, 404 (1952).
76. R. C. Tolman, *J. Chem. Phys.* **17**, 333 (1949).
77. J. G. Kirkwood and F. P. Buff, *J. Chem. Phys.* **17**, 338 (1949).
78. F. P. Buff, *J. Chem. Phys.* **23**, 419 (1955).
79. H. Reiss, *J. Chem. Phys.* **20**, 1216 (1952).
80. H. Reiss, private communication.
81. J. W. Cahn and J. E. Hilliard, *J. Chem. Phys.* **28**, 258 (1958).
82. E. W. Hart, *Phys. Rev.* **113**, 412 (1959).
83. J. W. Cahn, *J. Chem. Phys.* **30**, 1121 (1959).
84. J. W. Cahn and J. E. Hilliard, *J. Chem. Phys.* **31**, 688 (1959).
85. M. Volmer and H. Flood, *Z. phys. Chem.* **A170**, 273 (1934).
86. W. G. Courtney, *J. Chem. Phys.* **35**, 2249 (1961); **36**, 2018 (1962); **38**, 1448 (1963).
87. J. P. Hirth, *Ann. N.Y. Acad. Sci.* **101**, 805 (1963).
88. C. T. R. Wilson, *Phil. Trans. Roy. Soc.* (*London*) **192**, 403 (1899).
89. C. T. R. Wilson, *Phil. Trans. Roy. Soc.* (*London*) **193**, 289 (1899).
90. C. T. R. Wilson, *Phil. Mag.* **7**, 681 (1904).
91. C. F. Powell, *Proc. Roy. Soc.* (*London*) **A119**, 553 (1928).
92. J. Zeldovich, *Acta Physicochim* (Russ.) **18**, 1 (1943).
93. A. Kantrowitz, *J. Chem. Phys.* **19**, 1097 (1951).
94. R. F. Probstein, *J. Chem. Phys.* **19**, 619 (1951).
95. F. C. Collins, *Z. Elektrochem.* **59**, 404 (1955).
96. H. Wakeshima, *J. Chem. Phys.* **22**, 1614 (1954).
97. D. Turnbull, *Trans. AIME* **175**, 774 (1948).
98. H. L. Frisch, *J. Chem. Phys.* **27**, 90 (1957).
99. J. C. Fisher, J. H. Hollomon and D. Turnbull, *J. Appl. Phys.* **19**, 775 (1948).
100. G. Wulff, *Z. Krist.* **34**, 449 (1901).

101. C. Herring, in *Structure and Properties of Solid Surfaces*, edited by R. Gomer and C. S. Smith, pp. 5–72, University of Chicago Press (1953).
102. A. Dinghas, *Z. Krist.* **105**, 304 (1944).
103. I. N. Stranski and R. Kaischew, *Z. phys. Chem.* (**B**) **26**, 317 (1934).
104. I. N. Stranski and R. Kaischew, *Physik. Z.* **36**, 393 (1935).
105. A. S. Skapski, *J. Chem. Phys.* **16**, 386 (1948).
106. H. Flood, *Z. phys. Chem.* (**A**) **170**, 286 (1934).
107. J. J. Thomson, *Conduction of Electricity through Gases*, Cambridge Univ. Press (1906).
108. O. Glemser, *Z. Elektrochem.* **44**, 341 (1938).
109. G. M. Pound and V. K. LaMer, *J. Am. Chem. Soc.* **74**, 2323 (1952).
110. H. Reiss, *J. Chem. Phys.* **18**, 529 (1950).
111. N. H. Fletcher, *J. Chem. Phys.* **29**, 572 (1958).
112. C. T. R. Wilson, *Trans. Roy. Soc.* (*London*) **A189**, 265 (1897).
113. W. Makower, *Phil. Mag.* **5**, 226 (1903).
114. H. Flood, Doctoral Dissertation, Berlin (1933).
115. W. E. Hazen, *Phys. Rev.* **65**, 259 (1944).
116. M. Pollermann and I. N. Stranski, *Diskussionstagung der Bunsengesellschaft*, Berlin (1952).
117. C. T. R. Wilson, *Proc. Roy. Soc.* (*London*) **A142**, 88 (1933).
118. J. G. Wilson, *Principles of Cloud Chamber Technology*, Cambridge Univ. Press (1951).
119. R. von Helmholtz, *Wied. Ann.* **27**, 508 (1886).
120. C. T. O'Konski and W. I. Higuchi, *J. Phys. Chem.* **60**, 1598 (1956).
121. W. I. Higuchi and C. T. O'Konski, *J. Coll. Sci.* **15**, 14 (1960).
122. A. Langsdorf, *Ind. Eng. Chem.* **44**, 1298 (1952).
123. V. J. Schaefer, *Ind. Eng. Chem.* **44**, 1300 (1952).
124. F. J. M. Farley, *Proc. Roy. Soc.* (*London*) **A212**, 530 (1952).
125. G. M. Pound and V. K. LaMer, *J. Chem. Phys.* **19**, 506 (1951).
126. J. V. Clarke and W. H. Rodebush, Nucleation in Supersaturated Vapors, Ph.D. Thesis of J. V. Clarke, University of Illinois, Urbana, Illinois (1953).
127. M. H. Edwards and W. C. Woodbury, *Can. J. Phys.* **38**, 335 (1960).
128. A. Sander and G. Damkoehler, *Naturwissen.* **31**, 460 (1943).
129. L. A. Madonna, C. M. Sciulli, L. N. Canjar and G. M. Pound, *Proc. Phys. Soc.* (*London*) **78**, 1218 (1961).
130. B. J. Mason, *Adv. in Phys.* **7**, 221 (1958).
131. J. Maybank and B. J. Mason, *Proc. Phys. Soc.* (*London*) **74**, 11 (1959).
132. L. A. Madonna, C. M. Sciulli, L. N. Canjar and G. M. Pound, unpublished research.
133. T. S. Needels, Condensation and Crystallization in a Wilson Cloud Chamber, Ph.D. Thesis, Ohio State University, Columbus, Ohio (1950).
134. T. H. Laby, *Phil. Trans. Roy. Soc.* (*London*) **208**, 445 (1908).
135. I. Scharrer, *Ann. Phys.* **35**, 619 (1939).
136. N. N. Das Gupta and S. K. Ghosh, *Reviews of Modern Physics* **18**, 225 (1946).
137. A. J. Barnard and W. L. Mouton, *Nature* **182**, 1001 (1958).
138. S. Datz and E. H. Taylor, *Recent Research in Molecular Beams*, edited by I. Estermann, p. 157, Academic Press, New York (1959).
139. J. Frenkel, *Z. Phys.* **26**, 117 (1924).
140. C. A. Wert and C. Zener, *Phys. Rev.* **76**, 1169 (1949).
141. A. Dupre, *Theorie Mecanique de la Chaleur*, Paris, p. 2883 (1869).
142. G. M. Pound, M. T. Simnad and L. Yang, *J. Chem. Phys.* **22**, 1215 (1954).
143. R. H. Fowler and E. A. Guggenheim, *Statistical Thermodynamics*, Macmillan, New York (1939).
144. J. P. Hirth, *Acta Met.* **7**, 755 (1959).
145. K. L. Moazed and G. M. Pound, Thesis, Carnegie Institute of Technology (1960); *Trans. AIME*, in press.
146. D. Turnbull and B. Vonnegut, *Ind. Eng. Chem.* **44**, 1292 (1952).
147. J. H. van der Merwe, *Proc. Phys. Soc.* (*London*) **63A**, 616 (1950).
148. H. Brooks, *Metal Interfaces*, p. 20, Amer. Soc. Metals (1952).
149. N. H. Fletcher, *J. Chem. Phys.* **30**, 1476 (1959).
150. D. Turnbull, *J. Chem. Phys.* **18**, 198 (1950).

151. B. K. CHAKRAVERTY and G. M. POUND, in ref. 536.
152. G. TAMMANN and R. F. MEHL, *States of Aggregation*, D. Van Nostrand, New York (1925).
153. R. GOMER, *J. Chem. Phys.* **28**, 457 (1958).
154. M. HUMENIK, private communication.
155. K. L. MOAZED, private communication.
156. J. W. MITCHELL, private communication.
157. J. A. BECKER, *Advances in Catalysis*, Vol. 7, p. 141, Academic Press, New York (1955).
158. A. J. DEKKER, *Solid State Physics*, p. 229, Macmillan, London (1958).
159. R. D. GRETZ, private communication.
160. J. N. CHIRIGOS, M. T. SIMNAD and G. M. POUND, Thesis, Carnegie Institute of Technology (1957), to be published.
161. I. G. PTUSHINSKII, *Zhur. Tekh. Fiz.* **28**, 1402 (1958).
162. E. VON GOELER and E. LÜSCHER, private communication.
163. H. MAYER and H. GOHRE, *Z. Phys.* in press.
164. M. KNUDSEN, *Ann. Phys.* **50**, 472 (1916).
165. C. CHARITON and M. SEMMENOFF, *Z. Phys.* **25**, 287 (1924).
166. I. ESTERMANN, *Z. Phys.* **33**, 320 (1925); *Z. Elektrochem.* **31**, 441 (1925).
167. J. D. COCKCROFT, *Proc. Roy. Soc.* (*London*) **119A**, 293 (1923).
168. L. YANG, C. E. BIRCHENALL, G. M. POUND and M. T. SIMNAD, *Acta Met.* **2**, 462 (1954).
169. G. W. SEARS, *Acta Met.* **3**, 367 (1955).
170. S. J. HRUSKA, private communication.
171. T. N. RHODIN, *Disc. Faraday Soc.* **5**, 215 (1949).
172. J. F. ELLIOT and M. GLEISER, *Thermochemistry for Steelmaking*, p. 16, Addison-Wesley, Reading, Mass., and Pergamon Press, London (1960).
173. K. G. GUNTHER, *Z. Phys.* **149**, 538 (1957).
174. G. S. YEH and B. M. SIEGEL, to be published.
175. J. M. M. WELLS and B. M. SIEGEL, to be published.
176. A. F. FRAY and S. NIELSEN, *Brit. J. Appl. Phys.* **12**, 603 (1961).
177. H. WEGENER, *Z. Phys.* **140**, 465 (1955).
178. F. M. DEVIENNE, *Compt. rend.*, **231**, 740 (1950); **234**, 30 (1952); *J. phys. rad.* **13**, 53 (1952).
179. F. M. DEVIENNE, *J. phys. rad.* **14**, 257 (1953).
180. F. M. DEVIENNE, *Compt. rend.*, **238**, 2397 (1954).
181. F. M. DEVIENNE, *Mem. sci. phys.* **53**, 1 (1952); *Vacuum* **3**, 392 (1953).
182. P. GARIN and P. PRUGNE, *J. phys. rad.* **15**, 829 (1954).
183. L. YANG, M. T. SIMNAD and G. M. POUND, *Acta Met.* **2**, 470 (1954).
184. H. FRAUNFELDER, *Helv. Phys. Acta* **23**, 347 (1950).
185. H. WALTHER, *Z. ang. Phys.* **10**, 272 (1958).
186. S. WEXLER, *Rev. Mod. Phys.* **30**, 402 (1958).
187. J. D. LIVINGSTON, *Trans. AIME* **215**, 566 (1959).
188. H. SEIFERT, *Structure and Properties of Solid Surfaces*, edited by R. GOMER and C. S. SMITH, p. 318, University of Chicago Press (1953).
189. G. A. BASSETT, J. W. MENTER and D. W. PASHLEY, ref. 15, p. 11.
190. D. W. PASHLEY, *Adv. in Phys.* **5**, 173 (1956).
191. F. C. FRANK and J. G. VAN DER MERWE, *Proc. Roy. Soc.* (*London*) **A198**, 205, 215 (1949); **A200**, 125 (1949).
192. H. LEVINSTEIN, *J. Appl. Phys.* **20**, 306 (1949).
193. S. A. SEMILETOV, *Kristallografia* **1**, 542 (1956).
194. L. BRUCK, *Ann. Phys.* **26**, 233 (1936).
195. R. B. KEHOE, *Phil. Mag.* **2**, 455 (1957).
196. R. W. HOFFMANN, discussion to reference 157.
197. B. W. SLOOPE and C. O. TILLER, *J. Appl. Phys.* **32**, 1331 (1961); **33**, 3461 (1962).
198. L. G. SHULZ, *Acta Cryst.* **5**, 130 (1952).
199. J. W. MATTHEWS, *Phil. Mag.* **6**, 1347 (1961).
200. P. DELAVIGNETTE, J. TOURNIER and S. AMELINCKX, *Phil. Mag.* **6**, 1419 (1961).
201. G. W. BRYANT, J. HALLETT and B. J. MASON, *J. Phys. Chem. Sol.* **12**, 189 (1959).
202. N. N. SEMENOV, *Zhur. Russ. Fiz. Khim. Obshchestva* **62**, 33 (1930).

203. W. Buckel, ref. 15, p. 53.
204. A. Götzberger, *Z. Phys.* **142**, 182 (1955).
205. L. S. Palatnik and Y. F. Komnik, *Dokl. Akad. Nauk SSSR* **124**, 808 (1959).
206. E. N. Da C. Andrade and J. G. Martindale, *Phil. Trans. Roy. Soc. (London)* **235A**, 69 (1935).
207. F. I. Metz and R. A. Lad, *J. Phys. Chem.* **60**, 277 (1956).
208. J. J. Gilman and W. G. Johnston, *Dislocations and Mechanical Properties of Crystals*, edited by Fisher *et al.*, p. 116, John Wiley, New York (1957).
209. S. C. Mossop, *Proc. Phys. Soc. (London)* **69B**, 161 (1956).
210. E. M. Fournier D'Albe, *Quart. J. Roy. Met. Soc.* **75**, 1 (1949).
211. G. A. Bassett, *Phil. Mag.* **3**, 72 (1958).
212. R. A. Aziz and G. D. Scott, *Can. J. Phys.* **34**, 731 (1956).
213. L. Holland, *Can. J. Phys.* **35**, 697 (1957).
214. R. A. Aziz and G. D. Scott, *Can. J. Phys.* **35**, 985 (1957).
215. D. M. Evans and H. Wilman, *Acta Cryst.* **5**, 731 (1952).
216. N. T. Melnikova, E. D. Shukin and M. M. Umanski, *Zhur. Eksp. Teor. Fiz.* **22**, 775 (1952).
217. B. M. Siegel and C. C. Peterson, ref. 15, p. 97.
218. M. Polanyi and E. Wigner, *Z. phys. Chem.* **139A**, 439 (1928).
219. K. Herzfeld, *J. Chem. Phys.* **3**, 319 (1935).
220. A. E. Stearn and H. Eyring, *J. Chem. Phys.* **5**, 113 (1937).
221. K. Neumann, *Z. phys. Chem.* **16**, 196 (1950).
222. S. S. Penner, *J. Phys. Chem.* **52**, 367, 949 and 1262 (1948); **56**, 475 (1952); **65**, 702 (1961).
223. R. D. Shultz and A. O. Dekker, *J. Chem. Phys.* **23**, 2133 (1955).
224. E. M. Mortensen and H. Eyring, *J. Phys. Chem.* **64**, 847 (1960).
225. P. C. Carman, *Trans. Faraday Soc.* **44**, 529 (1948).
226. K. J. Laidler, *Chemical Kinetics*, McGraw-Hill, New York (1950).
227. M. Knudsen, *Ann. Phys.* **47**, 697 (1915).
228. K. Neumann and K. Schmoll, *Z. phys. Chem.* **2**, 215 (1954).
229. K. Neumann and E. Völker, *Z. phys. Chem.* **161**, 33 (1932).
230. M. Volmer and I. Estermann, *Z. Phys.* **7**, 1, 13 (1921).
231. R. B. Holden, R. Speiser and H. L. Johnston, *J. Am. Chem. Soc.* **70**, 3897 (1943).
232. W. Prüger, *Z. Phys.* **115**, 202 (1940).
233. M. Baranaev, *J. Phys. Chem. USSR* **13**, 1635 (1939).
234. G. Wyllie, *Proc. Roy. Soc. (London)* **197A**, 383 (1949).
235. H. Bucka, *Z. phys. Chem.* **195**, 260 (1950).
236. D. J. Trevoy, *Ind. Eng. Chem.* **45**, 2366 (1953).
237. K. C. D. Hickman, *Ind. Eng. Chem.* **46**, 1442 (1954).
238. K. C. D. Hickman and D. J. Trevoy, *Ind. Eng. Chem.* **44**, 1882 (1952).
239. R. Littlewood and E. Rideal, *Trans. Faraday Soc.* **52**, 1598 (1956).
240. J. Birks and R. S. Bradley, *Proc. Roy. Soc. (London)* **198A**, 226 (1949).
241. R. S. Bradley and G. C. S. Waghorn, *Proc. Roy. Soc. (London)* **206A**, 65 (1951).
242. R. S. Bradley and A. D. Shellard, *Proc. Roy. Soc. (London)* **198A**, 239 (1949).
243. K. C. D. Hickman and W. A. Torpey, *Ind. Eng. Chem.* **46**, 1446 (1954).
244. N. H. Fletcher, *Phil. Mag.* **7**, 255 (1962).
245. V. K. LaMer, *Retardation of Evaporation by Monolayers*, Academic Press, New York (1961).
246. J. Stefan, *Wiener Ber.* **98**, 1418 (1889).
247. H. Mache, *Z. Phys.* **107**, 310 (1937).
248. I. Langmuir, *Phys. Rev.* **12**, 368 (1918).
249. N. Fuchs, *Phys. Z. Sowjet* **6**, 225 (1934).
250. R. S. Bradley, M. G. Evans and R. W. Whytlaw-Gray, *Proc. Roy. Soc. (London)* **186A**, 368 (1946).
251. O. Knacke and I. N. Stranski, in *Progress in Metal Physics*, edited by B. Chalmers, Vol. 6, Pergamon Press, p. 202 (1956).
252. L. Monchick and H. Reiss, *J. Chem. Phys.* **22**, 831 (1954).
253. H. L. Frisch and F. C. Collins, *J. Chem. Phys.* **20**, 1797 (1952).
254. J. S. Kirkaldy, *Can. J. Phys.* **36**, 446 (1958).

255. R. SPEISER and J. W. SPRETNAK, *Vacuum Metallurgy*, Electrochemical Society, Boston, p. 155 (1955).
256. W. E. BELL and U. MERTEN, to be published.
257. N. GUDRIS and L. KULIKOWA, *Z. Phys.* **25**, 121 (1924).
258. F. C. FRANK, *Growth and Perfection of Crystals*, edited by DOREMUS *et al.*, p. 3, John Wiley, New York (1958).
259. F. C. FRANK, private communication.
260. W. K. BURTON, N. CABRERA and F. C. FRANK, *Phil. Trans. Roy. Soc.* (*London*) **A243**, 299 (1950).
261. J. W. CAHN, *Acta Met.* **8**, 554 (1960).
262. J. P. HIRTH and G. M. POUND, *J. Chem. Phys.* **26**, 1216 (1957).
263. W. K. BURTON and N. CABRERA, *Disc. Faraday Soc.* **5**, 40 (1949).
264. R. KAISCHEW, *Acta Phys. Acad. Sci. Hungar.* **8**, 19 (1957).
265. G. W. SEARS, *J. Chem. Phys.* **31**, 157 (1959).
266. W. T. READ, Jr., *Dislocations in Crystals*, McGraw-Hill, New York (1953).
267. D. McLEAN, *Grain Boundaries in Metals*, Clarendon Press, Oxford (1957).
268. W. W. MULLINS, *Acta Met.* **7**, 746 (1959).
269. O. KNACKE, I. N. STRANSKI and G. WOLFF, *Z. Elektrochem.* **56**, 476 (1952); *Z. physik. Chem.* **198**, 157 (1951).
270. Ref. 4, p. 27.
271. I. N. STRANSKI, *Z. phys. Chem.* **11**, 342 (1931).
272. Ref. 251, p. 224.
273. M. KNUDSEN, *Ann. Phys.* **29**, 179 (1909).
274. F. C. FRANK, ref. 258, p. 411.
275. M. J. LIGHTHILL and G. B. WHITHAM, *Proc. Roy. Soc.* (*London*) **229A**, 281, 317 (1955).
276. N. CABRERA and D. A. VERMILYEA, ref. 258, p. 393.
277. A. A. CHERNOV, *Dokl. Akad. Nauk. SSSR* **117**, 983 (1957).
278. W. W. WEBB, ref. 258, p. 230.
279. P. CHARSLEY and P. E. RUSH, *Phil. Mag.* **3**, 508 (1958).
280. P. B. PRICE, *Phil. Mag.* **5**, 417 (1960).
281. W. C. DASH, *J. Appl. Phys.* **29**, 736 (1958); ref. 258, p. 361.
282. G. W. SEARS, *Acta Met.* **3**, 361 (1955).
283. G. W. SEARS, *Acta Met.* **1**, 457 (1953).
284. T. R. HOGNESS, *J. Am. Chem. Soc.* **43**, 1621 (1921).
285. E. R. FUNK, H. UDIN and J. WULFF, *Trans. AIME* **191**, 1207 (1951).
286. H. SAMELSON, *J. Appl. Phys.* **32**, 309 (1961).
287. C. R. MORELOCK, *Acta Met.* **10**, 161 (1962).
288. P. B. PRICE, *Phil. Mag.* **5**, 473 (1960).
289. P. B. PRICE, *J. Appl. Phys.* **32**, 1747 (1961).
290. P. B. PRICE, private communication.
291. R. C. DEVRIES and G. W. SEARS, *J. Chem. Phys.* **31**, 1256 (1959).
292. G. W. SEARS, *J. Chem. Phys.* **24**, 868 (1956).
293. G. W. Sears, *J. Chem. Phys.* **27**, 1308 (1957).
294. *Handbook of Chemistry and Physics*, p. 1155, Chemical Rubber Publishing Company, Cleveland, Ohio (1950).
295. J. B. HUDSON and G. W. SEARS, *J. Chem. Phys.* **35**, 1509 (1961).
296. F. C. FRANK and M. B. IVES, *J. Appl. Phys.* **31**, 1996 (1960).
297. M. B. IVES, Ph.D. Thesis, University of Bristol (1960).
298. B. W. BATTERMAN, *J. Appl. Phys.* **28**, 1236 (1957).
299. M. B. IVES and J. P. HIRTH, *J. Chem. Phys.* **33**, 517 (1960).
300. G. W. SEARS, *J. Chem. Phys.* **25**, 154 (1956).
301. D. R. BRAME and T. EVANS, *Phil. Mag.* **3**, 971 (1958).
302. V. A. PHILLIPS, *Phil. Mag.* **5**, 571 (1960).
303. J. W. MATTHEWS, *Phil. Mag.* **4**, 1017 (1959).
304. K. BAHADUR and P. V. SASTRY, *Proc. Phys. Soc.* (*London*) **78**, 594 (1961).
305. I. N. STRANSKI, *Z. Elektrochem.* **35**, 393 (1929).
306. G. W. SEARS, *J. Chem. Phys.* **24**, 1045 (1958).
307. N. CABRERA, in *Surface Chemistry of Metals and Semiconductors*, edited by H. C. GATOS, p. 71, John Wiley, New York (1959).

308. A. A. CHERNOV, *Kristallografia* **1**, 119 (1956).
309. W. DITTMAR and K. NEUMANN, *Z. Elektrochem.* **61**, 70 (1957).
310. W. DITTMAR and K. NEUMANN, *Z. Elektrochem.* **63**, 738 (1959).
311. W. DITTMAR and K. NEUMANN, ref. 258, p. 121.
312. W. DITTMAR and K. NEUMANN, *Z. Elektrochem.* **64**, 297 (1960).
313. S. J. HRUSKA and J. P. HIRTH, *Z. Elektrochem.* **65**, 479 (1961).
314. R. SUHRMANN, ref. 310, p. 742.
315. G. W. SEARS, *J. Chem. Phys.* **33**, 1068 (1960).
316. G. W. SEARS, *J. Chem. Phys.* **29**, 979 (1958).
317. A. S. MICHAELS and A. R. COLVILLE, Jr., *J. Phys. Chem.* **64**, 13 (1960).
318. A. S. MICHAELS and F. W. TAUSCH, Jr., *J. Phys. Chem.* **65**, 1731 (1961).
319. A. S. MICHAELS, private communication.
320. A. A. CHERNOV, *Sov. Phys. Uspekhi* **4**, 116 (1961).
321. W. DEKEYSER and S. AMELINCKX, *Les Dislocations et la Croissance des Cristaux*, Masson, Paris (1956).
322. A. R. VERMA, *Crystal Growth and Dislocations*, Butterworths Scientific Publications, London (1953).
323. W. G. COURTNEY, *Amer. Rocket Soc. Journal* **31**, 751 (1961).
324. F. C. FRANK, *Disc. Faraday Soc.* **5**, 48, 67 (1949).
325. N. CABRERA and M. M. LEVINE, *Phil. Mag.* **1**, 450 (1956).
326. J. P. HIRTH and G. M. POUND, *Acta Met.* **5**, 649 (1957).
327. R. A. ORIANI, *J. Chem. Phys.* **18**, 575 (1950).
328. R. GOMER and C. S. SMITH, *Structure and Properties of Solid Surfaces*, University of Chicago Press (1953).
329. S. S. BRENNER and G. W. SEARS, *Acta Met.* **4**, 268 (1956).
330. F. C. FRANK, *Phil. Mag.* **42**, 1014 (1951).
331. A. J. FORTY, *Phil. Mag.* **43**, 377 (1952).
332. F. C. FRANK, *Physica* **15**, 131 (1949).
333. N. CABRERA, *J. Chem. Phys.* **21**, 1111 (1953).
334. A. R. LANG, *J. Appl. Phys.* **28**, 497 (1957).
335. N. CABRERA, ref. 328, p. 294.
336. M. LIFSCHITZ and A. A. CHERNOV, *Sov. Phys. Cryst.* **4**, 746 (1959).
337. W. W. WEBB, private communication.
338. F. C. FRANK, *Adv. in Phys.* **1**, 91 (1952).
339. J. C. FISHER, R. L. FULLMAN and G. W. SEARS, *Acta Met.* **2**, 344 (1954).
340. J. B. NEWKIRK and G. W. SEARS, *Acta Met.* **3**, 110 (1955).
341. G. EHRLICH, *Acta Met.* **3**, 201 (1955).
342. J. P. HIRTH and F. C. FRANK, *Phil. Mag.* **3**, 1110 (1958).
343. A. J. FORTY, private communication.
344. S. A. KITCHENER and R. F. STRICKLAND-CONSTABLE, *Proc. Roy. Soc.* (*London*) **A245**, 93 (1958).
345. R. S. BRADLEY and T. DRURY, *Trans. Faraday Soc.* **55**, 1848 (1959).
346. M. VOLMER and W. SCHÜLTZE, *Z. phys. Chem.* **156A**, 1 (1931).
347. J. E. MCNUTT and R. F. MEHL, *Trans. ASM* **50**, 1006 (1958).
348. G. G. LEMMLEIN, E. D. DUKOVA and A. A. CHERNOV, *Kristallografia* **2**, 428 (1957).
349. R. L. PARKER and L. M. KUSHNER, *J. Chem. Phys.* **35**, 1345 (1961).
350. A. A. CHERNOV and E. D. DUKOVA, *Sov. Phys. Cryst.* **5**, 627 (1961).
351. J. HALLETT, *Phil. Mag.* **6**, 1073 (1961).
352. M. I. KOZLOVSKII and G. G. LEMMLEIN, *Sov. Phys. Cryst.* **3**, 352 (1960).
353. R. KAISCHEW, E. BUDEWSKI and J. MALINOWSKI, *Z. phys. Chem.* **204**, 348 (1955).
354. F. M. DEVIENNE, *Vacuum* **3**, 392 (1953).
355. S. CHANDRA and G. D. SCOTT, *Can. J. Phys.* **36**, 1148 (1958).
356. R. A. RAPP, J. P. HIRTH and G. M. POUND, *Can. J. Phys.* **38**, 709 (1960).
357. F. HOCK and K. NEUMANN, *Z. phys. Chem.* **2**, 241 (1954).
358. R. N. HAWARD, *Trans. Faraday Soc.* **35**, 1401 (1939).
359. R. A. RAPP, J. P. HIRTH and G. M. POUND, *J. Chem. Phys.* **34**, 184 (1961).
360. W. A. CHUPKA, to be published.
361. M. KNUDSEN, *Ann. Phys.* **50**, 472 (1916).
362. J. B. NEWKIRK, *Acta Met.* **4**, 316 (1956).
363. R. L. PARKER, *J. Chem. Phys.* **37**, 1600 (1962); ref. 536.

364. H. STAHL, *J. Appl. Phys.* **20**, 1 (1949); H. STAHL and S. WAGENER, *Z. techn. Phys.* **27**, 280 (1943).
365. Ref. 251, p. 181.
366. I. N. STRANSKI and A. KOLB, *Naturwiss.* **33**, 220 (1946).
367. I. N. STRANSKI and K. BECKER, *Z. Naturforsch.* **2**, 173 (1947).
368. I. N. STRANSKI and G. WOLFF, *Z. Elektrochem.* **53**, 1 (1949).
369. A. J. FORTY, *Adv. in Phys.* **3**, 1 (1954).
370. L. J. GRIFFIN, *Phil. Mag.* **41**, 196 (1950).
371. A. R. VERMA, *Nature* **167**, 939 (1951); *Phil. Mag.* **43**, 441 (1952).
372. S. AMELINCKX, *Nature* **167**, 939 (1951).
373. S. AMELINCKX, *Phil. Mag.* **43**, 562 (1952).
374. A. J. FORTY, *Phil. Mag.* **43**, 481 (1952).
375. W. I. POLLOCK and R. F. MEHL, *Acta Met.* **3**, 213 (1955); *Trans. ASM* **51**, 162 (1959).
376. A. J. FORTY, *Phil. Mag.* **43**, 481, 949 (1952).
377. S. TANISAKI, *J. Phys. Soc.* (*Japan*) **11**, 620 (1956).
378. M. BRANDSTÄTER, *Z. Elektrochem.* **56**, 968 (1952).
379. F. C. FRANK and A. J. FORTY, *Proc. Roy. Soc.* (*London*) **A217**, 262 (1953).
380. E. D. DUKOVA, *Sov. Phys. Dokl.* **3**, 703 (1958).
381. A. J. FORTY, *Phil. Mag.* **43**, 72, 377 (1952).
382. J. B. NEWKIRK, *Acta Met.* **3**, 121 (1955).
383. I. M. DAWSON and V. VAND, *Proc. Roy. Soc.* (*London*) **206A**, 555 (1951).
384. I. M. DAWSON, *Proc. Roy. Soc.* (*London*) **214A**, 72 (1952).
385. P. M. REYNOLDS and A. R. VERMA, *Nature* **171**, 486 (1953).
386. N. G. ANDERSON and I. M. DAWSON, *Proc. Roy. Soc.* (*London*) **218A**, 255 (1953).
387. G. G. LEMMLEIN and E. D. DUKOVA, *Kristallografia* **1**, 477 (1956).
388. B. VLACH, *Sov. Phys. Cryst.* **6**, 519 (1962).
389. A. ACCARY and R. F. MEHL, *Compt. rend.* **244**, 2713 (1957).
390. A. KORNDORFFER, H. RAHBEK and F. SULTAN, *Phil. Mag.* **43**, 1301 (1952).
391. G. W. SEARS and R. V. COLEMAN, *J. Chem. Phys.* **25**, 635 (1956).
392. M. I. KOZLOVSKII, *Sov. Phys. Cryst.* **3**, 206 (1960).
393. K. NEUMANN and W. DITTMAR, *Naturwissen.* **42**, 510 (1955).
394. L. KOWARSKI, *J. Chem. Phys.* **32**, 303 (1935).
395. E. YODA, *J. Phys. Soc.* (*Japan*) **15**, 821 (1960).
396. F. R. N. NABARRO and P. J. JACKSON, ref. 258, p. 13.
397. R. V. COLEMAN and G. W. SEARS, *Acta Met.* **5**, 131 (1957).
398. R. V. COLEMAN and N. CABRERA, *J. Appl. Phys.* **28**, 1360 (1957).
399. T. KOBAYASHI, *Phil. Mag.* **6**, 1363 (1961).
400. J. D. ESHELBY, *J. Appl. Phys.* **24**, 176 (1953).
401. G. W. SEARS, *J. Chem. Phys.* **31**, 53 (1959).
402. G. W. SEARS, R. C. DEVRIES and C. HUFFINE, *J. Chem. Phys.* **34**, 2142 (1961).
403. A. J. MELMED and R. GOMER, *J. Chem. Phys.* **34**, 1802 (1961); **30**, 586 (1959).
404. R. L. PARKER and S. C. HARDY, *J. Chem. Phys.* **37**, 1606 (1962); ref. 536.
405. T. HOFFMAN, J. MAZUR, J. NIKLIBORC and J. RAFALOWICZ, *Brit. J. Appl. Phys.* **12**, 635, 342 (1961).
406. W. J. SPENCER and R. D. DRAGSDORF, *J. Appl. Phys.* **33**, 239 (1962).
407. A. J. FORTY, *Phil. Mag.* **5**, 787 (1960); **6**, 587 (1961).
408. A. J. FORTY, private communication.
409. N. CABRERA, *Semiconductor Surface Physics*, edited by R. H. KINGSTON, p. 327, University of Pennsylvania Press, Philadelphia (1957).
410. N. CABRERA, M. M. LEVINE and J. S. PLASKETH, *Phys. Rev.* **96**, 1153 (1954).
411. F. C. FRANK, *Acta Cryst.* **4**, 497 (1951).
412. D. A. VERMILYEA, *Acta Met.* **6**, 381 (1958).
413. I. LANGMUIR, *Phys. Rev.* **2**, 329 (1913).
414. C. L. MCCABE and E. BIRCHENALL, *Trans. AIME* **197**, 707 (1953).
415. I. V. KORNEV and E. Z. VINTAIKIN, *Dokl. Akad. Nauk. SSSR* **107**, 661 (1956).
416. G. WESSEL, *Z. Phys.* **130**, 539 (1951).
417. H. A. JONES, I. LANGMUIR and G. M. J. MACKAY, *Phys. Rev.* **30**, 201 (1927).
418. J. P. HIRTH and G. M. POUND, *Trans. AIME* **215**, 932 (1959).
419. P. HARTECK, *Z. phys. Chem.* **134**, 1 (1928).
420. H. N. HERSH, *J. Am. Chem. Soc.* **75**, 1529 (1953).

421. R. B. McClellan and R. Shuttleworth, *Z. Met.* **51**, 143 (1960).
422. A. L. Marshall, R. W. Dornte and F. J. Norton, *J. Am. Chem. Soc.* **59**, 1161 (1937).
423. J. W. Edwards, H. L. Johnston and W. E. Ditmars, *J. Am. Chem. Soc.* **75**, 2467 (1953).
424. R. Schuman and A. B. Garrett, *J. Am. Chem. Soc.* **66**, 442 (1942).
425. E. A. Gulbransen and K. E. Andrew, *J. Electrochem. Soc.* **97**, 383 (1950).
426. T. A. O'Donnell, *Austral. J. Chem.* **8**, 485 (1955).
427. A. C. Edgerton and F. V. Raliegh, *J. Chem. Soc.* **123**, 3024 (1923).
428. K. Bennewitz, *Ann. Phys.* **59**, 193 (1919).
429. C. L. McCabe, R. G. Hudson and H. W. Paxton, *Trans. AIME* **212**, 102 (1958).
430. V. D. Burkalov, *Phys. Met. and Met.* **5**, 72 (1957).
431. J. W. Edwards, H. L. Johnston and W. E. Ditmars, *J. Am. Chem. Soc.* **73**, 4729 (1951).
432. R. Speiser, H. L. Johnston and P. Blackburn, *J. Am. Chem. Soc.* **72**, 4142 (1950).
433. E. A. Gulbransen and K. E. Andrew, *J. Electrochem. Soc.* **99**, 402 (1952).
434. O. Knacke and R. Schmolke, *Z. Met.* **47**, 22 (1956).
435. R. S. Bradley and P. Volans, *Proc. Roy. Soc. (London)* **217A**, 508 (1933).
436. O. Knacke, R. Schmolke and I. N. Stranski, *Z. Krist.* **109**, 184 (1957).
437. R. S. Bradley, *Proc. Roy. Soc. (London)* **205A**, 553 (1951).
438. E. Rideal and P. M. Wiggins, *Proc. Roy. Soc. (London)* **210A**, 291 (1952).
439. G. M. Rothberg, M. Eisenstadt and P. Kusch, *J. Chem. Phys.* **30**, 517 (1959).
440. H. Kramers and S. Stemerding, *Appl. Sci. Res.* **3A**, 73 (1953).
441. K. Tschudin, *Helv. Phys. Acta* **19**, 91 (1946).
442. M. Baranaev, *J. Phys. Chem. USSR* **20**, 399 (1946).
443. F. Metzger and E. Miescher, *Nature* **142**, 572 (1938).
444. F. Metzger, *Helv. Phys. Acta* **16**, 323 (1943).
445. H. W. Melville and S. C. Gray, *Trans. Faraday Soc.* **32**, 271, 1026 (1936).
446. J. S. Kane, Thesis, University of California (1956).
447. J. S. Kane and J. H. Reynolds, *J. Chem. Phys.* **25**, 342 (1956).
448. H. Spingler, *Z. phys. Chem.* **52B**, 90 (1942).
449. G. W. Sears and L. Navais, *J. Chem. Phys.* **30**, 1111 (1959).
450. L. H. Dreger, V. V. Dadape and J. L. Margrave, to be published.
451. R. J. Thorn and G. H. Winslow, *J. Chem. Phys.* **26**, 186 (1957).
452. W. W. Mullins, *Phil. Mag.* **6**, 1313 (1961).
453. W. L. Winterbottom and J. P. Hirth, *J. Chem. Phys.* **37**, 784 (1962).
454. L. G. Carpenter and W. N. Mair, *Trans. Faraday Soc.* **55**, 1924 (1959).
455. U. Ernas and P. G. Shewmon, unpublished research.
456. M. G. Inghram and J. Drowart, in *High Temperature Technology*, p. 219, McGraw-Hill, New York (1960).
457. V. J. Clancy, *Nature* **166**, 275 (1950).
458. R. S. Bradley, *Proc. Roy. Soc. (London)* **217A**, 524 (1953).
459. P. Clausing, *Ann. Phys.* **12**, 961 (1932).
460. R. C. Miller and P. Kusch, *J. Chem. Phys.* **25**, 860 (1956); **27**, 981 (1957).
461. E. Miescher, *Helv. Phys. Acta* **14**, 507 (1941).
462. L. Pauling and M. Simonetta, *J. Chem. Phys.* **20**, 29 (1952).
463. W. A. Chupka and M. G. Inghram, *J. Phys. Chem.* **59**, 100 (1955).
464. G. DeMaria, J. Drowart and M. G. Inghram, *J. Chem. Phys.* **30**, 827 (1959).
465. E. D. Dukova, *Sov. Phys. Cryst.* **6**, 357 (1961).
466. E. Votava and S. Amelinckx, *Naturwissen.* **41**, 422 (1954).
467. E. Votava, *Z. Met.* **47**, 309 (1956).
468. H. Bethge and O. Schaffer, *Naturwissen.* **41**, 573 (1954).
469. E. Votava, A. Berghezan and R. H. Gillette, *Naturwissen.* **44**, 372 (1957).
470. E. Votava and A. Berghezan, *Acta Met.* **7**, 392 (1959).
471. H. Suzuki, *J. Phys. Soc. (Japan)* **10**, 981 (1955).
472. F. W. Young, Jr., *J. Appl. Phys.* **27**, 554 (1956).
473. M. J. Fraser, D. Caplan and A. A. Burr, *Acta Met.* **4**, 186 (1956).
474. W. Rosenhain and D. Ewen, *J. Inst. Met.* **8**, 149 (1912).
475. J. A. Leroux and E. Raub, *Z. anorg. Chem.* **188**, 205 (1930).
476. B. Chalmers, R. King and R. Shuttleworth, *Proc. Roy. Soc. (London)* **193A**, 465 (1948).

477. E. D. Hondros and A. J. W. Moore, *Acta Met.* **8**, 647 (1960).
478. F. W. Young, Jr. and A. T. Gwathmey, *J. Appl. Phys.* **31**, 225 (1960).
479. W. G. Johnston, in *Progress in Ceramic Science*, Vol. 2, edited by J. E. Burke, Pergamon Press (1962).
480. J. C. Danko and A. J. Griest, *Trans. AIME* **206**, 515 (1956).
481. E. Kern and H. Pick, *Z. Phys.* **134**, 610 (1953).
482. P. Evans, *Acta Met.* **5**, 342 (1957).
483. A. A. Hendrickson and E. S. Machlin, *Acta Met.* **3**, 64 (1955).
484. J. J. Gilman, W. J. Johnston and G. W. Sears, *J. Appl. Phys.* **29**, 747 (1958).
485. W. C. Dash, *J. Appl. Phys.* **27**, 1193 (1956).
486. L. Vassamillet and J. P. Hirth, *J. Appl. Phys.* **29**, 595 (1958).
487. H. Lambot, L. Vassamillet and J. Dejace, *Acta Met.* **1**, 711 (1953).
488. W. Doering, *Z. phys. Chem.* **B36**, 371 (1937).
489. W. Doering, *Z. phys. Chem.* **B38**, 292 (1938).
490. L. Bernath, *Ind. Eng. Chem.* **44**, 1310 (1952).
491. K. L. Wismer, *J. Phys. Chem.* **26**, 301 (1922).
492. F. B. Kenrick, C. S. Gilbert and K. L. Wismer, *J. Phys. Chem.* **28**, 1297 (1924).
493. H. N. V. Temperley, *Proc. Phys. Soc.* (*London*) **59**, 119 (1947).
494. J. C. Fisher, *J. Appl. Phys.* **19**, 1062 (1948).
495. G. M. Lewis, *Proc. Phys. Soc.* (*London*) **78**, 133 (1961).
496. M. Bertholet, *Ann. Chim.* **30**, 232 (1850).
497. D. A. Glaser, *Phys. Rev.* **87**, 665 (1952).
498. D. A. Glaser, *Phys. Rev.* **91**, 762 (1953).
499. B. Alfredsson and T. Johansson, *Arkiv för Fysik* **19**, 383 (1961).
500. T. Johansson, *Arkiv för Fysik* **19**, 397 (1961).
501. H. Wakeshima, *J. Phys. Soc.* (*Japan*) **16**, 6 (1961).
502. J. Meyer, *Zur Kenntnis des negativen Druckes in Flüssigkeiten*, W. Knapp, Halle (1911).
503. R. S. Vincent, *Proc. Phys. Soc.* (*London*) **53**, 126 (1941).
504. A. F. Scott and G. M. Pound, *J. Chem. Phys.* **9**, 726 (1941).
505. L. J. Briggs, *J. Appl. Phys.* **21**, 721 (1950).
506. L. J. Briggs, *Science* **112**, 427 (1950).
507. L. J. Briggs, *Science* **113**, 483 (1951).
508. L. J. Briggs, *J. Chem. Phys.* **19**, 970 (1951).
509. L. J. Briggs, *J. Appl. Phys.* **24**, 488 (1953).
510. R. S. Barnes, *Proc. Phys. Soc.* (*London*) **65B**, 512 (1952).
511. R. W. Balluffi and B. Alexander, *J. Appl. Phys.* **23**, 937 (1952).
512. J. N. Greenwood, D. R. Miller and J. W. Suiter, *Acta Met.* **2**, 250 (1959).
513. P. B. Hirsch, J. Silcox, R. E. Smallman and K. H. Westmacott, *Phil. Mag.* **3**, 897 (1958).
514. P. B. Hirsch and J. Silcox, ref. 258, p. 262.
515. F. Seitz, *Phys. Rev.* **79**, 890 (1950).
516. F. C. Frank, *Carnegie Institute of Technology Symposium on the Plastic Deformation of Crystalline Solids*, p. 89, Office of Naval Research (1950).
517. D. Wilsdorf, *Phil. Mag.* **3**, 125 (1958).
518. E. Teghtsoonian and B. Chalmers, *Can. J. Phys.* **29**, 370 (1951); **30**, 388 (1952).
519. G. Schoeck and W. A. Tiller, *Phil. Mag.* **5**, 43 (1960).
520. R. Resnick and L. Seigle, *Trans. AIME* **209**, 87 (1957).
521. W. M. Lomer, in *Progress in Metal Physics*, Vol. 8, p. 255, Pergamon Press (1959).
522. H. Kimura, R. Maddin and D. Kuhlmann-Wilsdorf, *Acta Met.* **7**, 145, 154 (1959).
523. J. E. Bauerle and J. S. Koehler, *Phys. Rev.* **107**, 1493 (1957).
524. J. S. Koehler, F. Seitz and J. E. Bauerle, *Phys. Rev.* **107**, 1499 (1957).
525. F. H. Horn, private communication.
526. V. K. LaMer, *Ind. and Eng. Chem.* **44**, 1270 (1952).
527. V. K. LaMer and D. Sinclair, *Chem. Revs.* **44**, 245 (1949).
528. R. D. Gretz and G. M. Pound, *Proc. International Symposium on Condensation and Evaporation of Solids*, Dayton, September 1962.
529. V. K. LaMer and Ruth Gruen, *Trans. Faraday Soc.* **48**, 410 (1952).
530. I. N. Stranski and W. Hirschwald, *Proc. International Symposium on Condensation and Evaporation of Solids*, Dayton, September 1962.

531. D. Walton, *Phil. Mag.* **7**, 1671 (1962).
532. D. Walton, *J. Chem. Phys.* **37**, 1282 (1962).
533. T. N. Rhodin and D. Walton, in reference 535.
534. J. Mayer and M. G. Mayer, *Statistical Mechanics*, John Wiley, New York (1940).
535. *ASM-AIME Symposium on Surfaces; Structure, Energetics and Kinetics*, ASM, Cleveland (1963).
536. *Proceedings of the Dayton International Conference on Condensation and Evaporation*, Gordon and Breach, New York (1963).
537. Conference on Clean Surfaces, *Ann. N.Y. Acad. of Sci.*, **101**, (1963).
538. *Conference on Epitaxial Monocrystalline Thin Films*, Philco Research Laboratory, Blue Bell, Pennsylvania (1963).
539. S. J. Hruska, Ph.D. Thesis, Carnegie Institute of Technology, 1963, to be published.

AUTHOR INDEX

SUBJECT INDEX

CONTENTS OF PREVIOUS VOLUMES IN THE SERIES

CONTENTS OF PREVIOUS VOLUMES IN THE SERIES

VOLUME 5

VOLUME 6

VOLUME 7

VOLUME 8

VOLUME 9

VOLUME 10